Edexcel A2 | UNIT 4

Chemistry

Rates, Equilibria and Further
Organic Chemistry

George Facer

Philip Allan Updates, an imprint of Hodder Education, an Hachette UK Company, Market Place, Deddington, Oxfordshire OX15 0SE

Orders

Bookpoint Ltd, 130 Milton Park, Abingdon, Oxfordshire, OX14 4SB
tel: 01235 827720
fax: 01235 400454
e-mail: uk.orders@bookpoint.co.uk
Lines are open 9.00 a.m.–5.00 p.m., Monday to Saturday, with a 24-hour message answering service. You can also order through the Philip Allan Updates website: www.philipallan.co.uk

© Philip Allan Updates 2009

ISBN 978-0-340-94948-1

First printed 2009
Impression number   5 4 3 2
Year 2014 2013 2012 2011 2010 2009

This guide has been written specifically to support students preparing for the Edexcel A2 Chemistry Unit 4 examination. The content has been neither approved nor endorsed by Edexcel and remains the sole responsibility of the author.

Printed by MPG Books, Bodmin

Hachette UK's policy is to use papers that are natural, renewable and recyclable products and made from wood grown in sustainable forests. The logging and manufacturing processes are expected to conform to the environmental regulations of the country of origin.

P01348

Contents

Introduction

■ ■ ■

Content Guidance

■ ■ ■

Questions and Answers

Introduction

About this guide

This unit guide is one of a series covering the Edexcel specification for AS and A2 chemistry. It offers advice for the effective study of **Unit 4: Rates, Equilibria and Further Organic Chemistry**. Its aim is to help you *understand* the chemistry — it is not intended as a shopping list, enabling you to cram for an examination.

The guide has three sections.
- **Introduction** — this provides guidance on study and revision, together with advice on approaches and techniques to ensure you answer the examination questions in the best way that you can.
- **Content Guidance** — this section is not intended to be a textbook. It offers guide-lines on the main features of the content of Unit 4, together with particular advice on making study more productive.
- **Questions and Answers** — this shows you the sort of questions you can expect in the unit test. Answers are provided; in some cases, distinction is made between responses that might have been given by a grade-A candidate and typical errors that are often made. Careful consideration of these will improve your answers and, more importantly, will improve your understanding of the chemistry involved.

The effective understanding of chemistry requires time. No one suggests it is an easy subject, but even those who find it difficult can overcome their problems by the proper investment of time.

To understand the chemistry, you have to make links between the various topics. The subject is coherent; it is not a collection of discrete modules. These links come only with experience, which means time spent thinking about chemistry, working with it and solving chemical problems. Time produces fluency with the ideas. If you have that, together with good technique, the examination will look after itself.

The specification

The specification states the chemistry that can be examined in the unit tests and describes the format of those tests. This is not necessarily the same as what teachers might choose to teach or what you might choose to learn.

The purpose of this book is to help you with Unit Test 4, but don't forget that what you are doing is learning *chemistry*. The specification can be obtained from Edexcel, either as a printed document or from the web at **www.edexcel.org.uk**.

The unit test

The Unit 4 test lasts 1 hour 40 minutes and has three sections:
- Section A — about 20 multiple-choice questions, each with a choice of four answers.
- Section B — a mixture of short-answer and extended-answer questions. This section looks similar to the papers in the previous modular tests, where the more extended writing is worth about 4 or 5 marks.
- Section C — contains one or more questions that require analysis of data. These data might be on the question paper but you are also required to find the necessary information from the Edexcel data booklet.

Command terms

Examiners use certain words that require you to respond in a specific way. You must distinguish between these terms and understand exactly what each requires you to do.
- **Define** — give a simple definition without any explanation.
- **Identify** — give the name or formula of the substance.
- **State** — no explanation is required (nor should you give one).
- **Deduce** — use the information supplied in the question to work out your answer.
- **Suggest** — use your knowledge and understanding of similar substances or those with the same functional groups to work out the answer.
- **Compare** — make a statement about *both* substances being compared.
- **Explain** — use chemical theories or principles to say why a particular property is as it is.
- **Predict** — say what you think will happen on the basis of the principles that you have learned.

Calculations

You must show your working in order to score full marks. Be careful about significant figures. If a question does not specify the number of significant figures required, give your answer to *three significant figures* or to two decimal places for pH calculations.

Organic formulae

- **Structural formula** — you must give a structure that is unambiguous. For instance, $CH_3CH_2CH_2OH$ is acceptable, but C_3H_7OH could be either propan-1-ol or propan-2-ol and so is not acceptable. If a compound has a double bond, then this should be shown in the structural formula.
- **Displayed or full structural formula** — you must show all the *atoms* and all the *bonds*. 'Sticks' instead of hydrogen atoms will lose marks.

- **Shape** — if the molecule or ion is pyramidal, tetrahedral or octahedral you must make sure that your diagram looks three-dimensional. To do this, use wedges and dashes. Draw optical isomers as mirror images of each other. Geometric isomers must be drawn with bond angles of 120°. Make sure that the *bonds go to the correct atoms* — for example, the oxygen in an –OH group or the carbon in $-CH_3$ and –COOH groups.

Points to watch

- **Stable** — if you use this word, you must qualify it. For example: 'stable to heat'; 'the reaction is thermodynamically stable'; 'the reaction is kinetically stable'; or 'a secondary carbocation intermediate is more stable than a primary carbocation'.
- **Reagents** — if you are asked to identify a reagent, you must give its *full* name or formula. Phrases such as 'acidified dichromate(VI)' will not score full marks. You must give the reagent's full name, for example 'potassium dichromate(VI)'.
- **Conditions** — don't use abbreviations such as 'hur'.
- **Atoms, molecules and ions** — don't use these words randomly. Ionic compounds contain ions, not molecules.
- **Rules** — don't use rules such as Markovnikov or Le Chatelier to *explain*. However, they can be used to predict.
- **Melting and boiling** — when a molecular covalent substance (such as water) is melted or boiled, *covalent* bonds are *not* broken. So melting and boiling points are connected with the type and strength of *intermolecular* forces. When an ionic substance is melted, the ionic bonds are *not* broken — the substance is still ionic. The ions gain enough energy to separate.

Learning to learn

Learning is not instinctive — you have to develop suitable techniques to make good use of your time. In particular, chemistry has peculiar difficulties that need to be understood if your studies are to be effective from the start.

Planning

Busy people do not achieve what they do haphazardly. They plan — so that if they are working they mean to be working, and if they are watching television they have planned to do so. Planning is essential. You must know what you have to do each day and each week and set aside time to do it.

Be realistic in your planning. You cannot work all the time, so you must build in time for recreation and family responsibilities.

Targets

When devising your plan, have a target for each study period. This might be a particular section of the specification, or it might be rearranging of information from text into pictures, or the construction of a flowchart relating all the organic reactions you need to know. Whatever it is, be determined to master your target material before you leave it.

Reading chemistry textbooks

Chemistry textbooks are a valuable resource, not only for finding out the information for your homework but also to help you understand concepts of which you are unsure. They need to be read carefully, with a pen and paper to hand for jotting down things as you go — for example, making notes, writing equations, doing calculations and drawing diagrams. Reading and revising are *active* processes which require concentration. Looking vaguely at the pages is a waste of time. In order to become fluent and confident in chemistry, you need to master detail.

Chemical equations
Equations are quantitative, concise and internationally understood.

When you write an equation, check that:
• you have thought of the *type* of reaction occurring — for example, is it neutralisation, addition or disproportionation?
• you have written the correct formulae for all the substances
• your equation balances both for the numbers of atoms of each element and for charge
• you have not written something silly, such as having a strong acid as a product when one of the reactants is an alkali
• you have included *state symbols* in all thermochemical equations and otherwise if they have been asked for

Graphs
Graphs give a lot of information and they must be understood in detail rather than as a general impression. Take time over them. Note what the axes are, the units, the shape of the graph and what the shape means in chemical terms. Think about what could be calculated from the graph. Note if the graph flattens off and what that means. This is especially important in kinetics.

When drawing a graph, do not join up the points — draw a smooth line (straight or curved) as near as possible to all the points. However, if you are plotting a list, such as the first ionisation energies of the elements, you do join up the points.

Tables
These are a means of displaying a lot of information. You need to be aware of the table headings and the units of numerical entries. Take time over them. What trends can be seen? How do these relate to chemical properties? Sometimes it can be useful to convert tables of data into graphs. Make sure that you use all the data when answering an examination question.

Diagrams

Diagrams of apparatus should be drawn in section. When you see them, copy them and ask yourself why the apparatus has the features it has. What is the difference between a distillation and a reflux apparatus, for example? When you do practical work, examine each piece of the apparatus closely so that you know both its form and function.

Calculations

Calculations are not normally structured in A2 as they were in AS. Therefore, you need to *plan* the procedure for turning the data into an answer.

- Set your calculations out fully, making it clear what you are calculating at each step. Don't round figures up or down during a calculation. Either keep all the numbers on your calculator or write any intermediate answers to four significant figures.
- If you have time, check the accuracy of each step by recalculating it. It is so easy to enter a wrong number into your calculator or to calculate a molar mass incorrectly.
- Finally, check that you have the correct *units* in your answer and that you have given it to an appropriate number of *significant figures* — if in doubt, give it to three or, for pH calculations, to two decimal places.

Notes

Most students keep notes of some sort. Notes can take many forms: they might be permanent or temporary; they might be lists, diagrams or flowcharts. You have to develop your own styles. For example, notes that are largely words can often be recast into charts or pictures and this is useful for imprinting the material. The more you rework the material, the clearer it will become.

Whatever form your notes take, they must be organised. Notes that are not indexed or filed properly are useless, as are notes written at enormous length and those written so cryptically that they are unintelligible a month later.

Writing

There is some requirement for extended writing in Unit Test 4. You need to be able to write concisely and accurately. This requires you to marshal your thoughts properly and needs to be practised during your ordinary learning.

For experimental plans, it is a good idea to write your answer as a series of bullet points. There are no marks specifically for 'communication skills', but if you are not able to communicate your ideas clearly and accurately, you will not score full marks. The space available for an answer is a poor guide to the amount that you have to write — handwriting sizes differ hugely, as does the ability to write succinctly. Filling the space does not necessarily mean you have answered the question. The mark allocation suggests the number of points to be made, not the amount of writing needed.

Approaching the unit test

The unit test is designed to allow you to show the examiner what you know. Answering questions successfully is not only a matter of knowing the chemistry but is also a matter of technique.

Revision

- Start your revision in plenty of time. Make a list of what you need to do, emphasising the topics that you find most difficult — and draw up a detailed revision plan. Work back from the examination date, ideally leaving an entire week free from fresh revision before that date. Be realistic in your revision plan and then add 25% to the timings because everything takes longer than you think.
- When revising, make a note of difficulties and ask your teacher about them. If you do not make these notes you will forget to ask.
- Make use of past papers. Similar questions are regularly asked, so if you work through as many past papers and answers as possible, you will be in a strong position to obtain a top grade.
- When you use the Questions and Answers section of this guide, make a determined effort to write *your* answers *before* looking at the sample answers and examiner's comments.

The exam

- Read the question. Questions usually change from one examination to the next. A question that looks the same, at a cursory glance, to one that you have seen before usually has significant differences when read carefully. Needless to say, candidates do not receive credit for writing answers to their own questions.
- Be aware of the number of marks available for a question. This is an excellent pointer to the number of things you need to say.
- Do not repeat the question in your answer. The danger is that you fill up the space available and think that you have answered the question, when in reality some or maybe all of the real points have been ignored.
- Look for words in **bold** in a question and make sure that you have answered the question fully in terms of those words or phrases. For example, if the question asks you to define **standard entropy** of a substance, make sure that you explain the meaning of entropy as well as what the standard conditions are.
- Be careful in negative multiple-choice questions. The word NOT will be in capital letters.
- Questions in Unit Test 4 will often involve substances or situations that are new to you. This is deliberate and is what makes these questions synoptic. Don't be put off by large organic molecules. They are nothing more than a collection of functional groups which, you may assume, react independently of each other.

Unit Test 4 has three assessment objectives:

- AO1 is 'knowledge with understanding of science and How Science Works' and makes up 26.5% of the test. You should be able to:
  - recognise, recall and show understanding of specific chemical facts, principles, concepts, practical techniques and terminology
  - select, organise and communicate information clearly and logically, using specialist vocabulary where appropriate
- AO2 is 'application of knowledge and understanding of science and How Science Works' and makes up 47% of Unit Test 4. You should be able to:
  - analyse and evaluate scientific knowledge and processes
  - apply scientific knowledge and processes to unfamiliar situations including those related to issues
  - assess the validity, reliability and credibility of scientific information
- AO3 is 'How Science Works' and makes up 26.5% of the test. You should be able to:
  - describe ethical, safe and skilful practical techniques and processes, selecting appropriate qualitative and quantitative methods
  - analyse, interpret, explain and evaluate the methodology, results and impact of experimental and investigative activities

Content Guidance

This section is a guide to the content of **Unit 4: Rates, Equilibria and Further Organic Chemistry**. It does not constitute a textbook for Unit 4 material.

The main areas of this unit are:
- How fast? Rates
- How far? Entropy
- Equilibria
- Applications of rates and equilibrium
- Acid–base equilibria
- Further organic chemistry
- Spectroscopy and chromatography

How fast? Rates

You are expected to know the kinetics studied in Unit 2. In particular, you should have a thorough understanding of collision theory, including the Maxwell–Boltzmann distribution of molecular energies.

AS chemistry

Collision theory

The following are important factors:
- **Collision frequency** — how often the molecules collide in a given time.
- **Collision energy** — particles must collide with enough kinetic energy to cause reaction. The minimum energy that two molecules must have on collision in order to react is called the **activation energy, E_a**.
- **Orientation** on collision — no reaction will occur if the OH$^-$ ion collides with the CH$_3$ group in the nucleophilic substitution reaction between, for example, chloroethane and hydroxide ions. The collision must occur with the carbon carrying the chlorine.

Maxwell–Boltzmann distribution

The molecules in a gas or a solution have a wide range of kinetic energies. This is shown graphically in the Maxwell–Boltzmann distribution at two temperatures T_{cold} and T_{hot}.

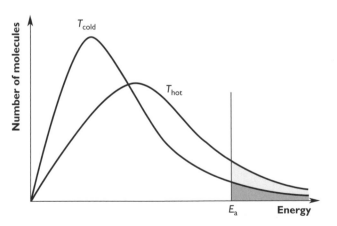

Note that both graphs start at the origin, but neither touches the x-axis at high energies. The difference between the two curves is that the curve for T_{hot} has:
- a lower modal value (the peak)
- a modal value to the right of that of T_{cold}

Effect of temperature on rate

The *area* under the curve to the *right* of the activation energy (shaded on the above graphs) is a measure of the number of molecules that have enough energy to react on collision. This area is greater on the T_{hot} graph compared with the T_{cold} graph. Therefore, the number of collisions between molecules with enough energy to react is greater. This means that a greater *proportion* of the collisions results in reaction, so the rate of reaction is faster at the higher temperature.

Effect of catalyst on rate

A catalyst causes the reaction to proceed by an *alternative route* that has a *lower* activation energy.

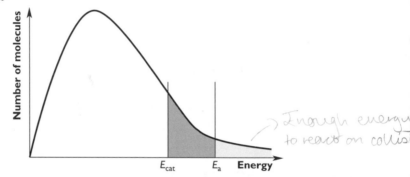

On the graph above, the value of E_{cat} is less than that of E_a, so the area under the curve to the right of E_{cat} is greater than the area to the right of E_a. This means that a greater *proportion* of molecules has the required lower activation energy on collision, allowing reaction to occur.

Effect of concentration and pressure on rate

An increase in the concentration of a reactant in solution, or in the pressure for gaseous reactions, increases the *frequency* of collisions. Therefore, the reaction rate increases.

Kinetic stability

A reaction mixture is kinetically stable if the activation energy is so high that the reaction is too slow to be observed at room temperature.

A2 chemistry

Definitions

- The **rate of reaction** is the amount by which the concentration changes in a given time. Its units are mol dm^{-3} s^{-1}. To determine reaction rate, the concentration of a reactant or product is measured at regular time intervals and a graph of concentration against time is plotted.

- The *slope (gradient)* of this line at a given point is the rate of reaction at that point.
- The slope is calculated by dividing the *change* in concentration by the time difference (see page 17).
- The **initial rate** of reaction is the rate of reaction at the time when the reactants are mixed. Its value can be estimated by measuring the concentration of a reactant at the start and after a short period of time. If the concentration has not changed by more than 10%, the initial rate is approximately equal to:

$$\frac{\text{change in concentration}}{\text{time}}$$

- The **rate equation** connects the rate of a reaction with the concentration of the reactants. For a reaction:

    xA + yB $\longrightarrow$ products

    the rate equation is:

    rate = $k[\text{A}]^m[\text{B}]^n$

    where m and n may or may not be the same as x and y. Note that if the reaction is carried out in the presence of a **homogeneous** catalyst, the concentration of the catalyst will appear in the rate equation. ⇥ same state from reactants
- The **order of reaction** is the sum of the powers to which the *concentrations* of the reactants are raised in the rate equation. In the example above, the order is $(m + n)$.
- The **partial order** with respect to one reactant is the power to which its concentration is raised in the rate equation. In the example above, the partial order with respect to A is m.
- The **rate constant**, k, is the constant of proportionality that connects the rate of the reaction with the concentration of the reactants, as shown in the above rate equation. Its value alters with temperature. A reaction with a *large* activation energy has a *low* value of k.
- The **half-life** of a reaction is the time taken for the concentration of a reagent to halve. This means that half of the reagent has reacted. The half-life of a first-order reaction is constant. The half-life of a second-order reaction increases as the concentration falls.
- In a multi-step reaction, the **rate-determining step** is the slowest step and controls the overall rate of the reaction. The reagents that appear in the rate equation are those that are involved in the rate-determining step and any previous steps. Any substance that only appears in the mechanism *after* the rate-determining step will not appear in the rate equation.
- The **activation energy**, E_a or E_{act}, of a reaction is the minimum energy that colliding molecules must have in order for that collision to result in a reaction.
- Catalysts can be described as either **heterogeneous** or **homogeneous**. A homogeneous catalyst is in the *same phase* as the reactants. An example is aqueous iron(II) ions in the oxidation of aqueous iodide ions by aqueous persulfate ions. A heterogeneous catalyst is in a *different* phase from the reactants. An example is the solid

iron catalyst in the reaction between gaseous nitrogen and hydrogen to produce ammonia (the Haber process).

Experimental methods

All the methods described below must be carried out at *constant temperature*, preferably using a thermostatically controlled water bath. However, the laboratory itself is often considered as a temperature-controlled environment, but this can cause considerable inaccuracy if the reaction is significantly exothermic.

Sampling

- Mix the reactants, stir and start timing.
- Remove samples at regular intervals and quench the reaction by adding the samples to iced water or a substance that will react with the catalyst.
- Titrate samples against a suitable solution.

This method can be used when a reactant or product is:

- an acid — titrate against an alkali
- a base — titrate against an acid
- iodine — titrate against sodium thiosulfate solution, using starch as an indicator

Continuous monitoring

- If a gas is produced, *either* measure the volume of gas at regular intervals (collecting it over water in a measuring cylinder or using a gas syringe) *or* carry out the reaction on a top pan balance and measure the loss in mass at regular intervals.
- If a reactant or product is coloured, follow the reaction in a colorimeter. The intensity of the colour is a measure of the concentration.
- If a reactant is a single optical isomer, follow the reaction using a polarimeter. The angle of rotation depends on the concentration.

Note: both a colorimeter and a polarimeter must first be standardised using a solution of known concentration. The use of a pH meter to follow the change in acidity is inaccurate.

The fixed point method

Some reactions can be followed by measuring the time taken to produce a fixed amount of a detectable product. The experiment is then repeated, changing the concentration of one of the reactants. In this type of experiment, the rate is calculated from the *reciprocal* of the time, i.e. 1/time is a measure of the rate.

Note: this method is only accurate if 10% or less of the reactant has reacted when the fixed amount of product is observed.

Some examples are:

- The reaction between dilute acid and aqueous sodium thiosulfate. The reactants are mixed in a beaker standing on a tile marked with a large X. The time taken for enough sulfur to be precipitated to hide the X is measured.

- The reaction between dilute acid and a metal such as magnesium or a carbonate such as calcium carbonate. The time taken for a fixed volume of gas is measured and the rate is proportional to 1/time.
- If the acid is in considerable excess, the time for all the magnesium to disappear is measured. As long as the acid's concentration has not changed by more than 10% during this time, 1/time is approximately proportional to the rate.

Clock methods

The reaction between an oxidising agent, such as hydrogen peroxide, and iodide ions in acid solution can be followed by this method.

A known amount of hydrogen peroxide is added to a solution of acid, iodide ions, starch and sodium thiosulfate and a clock started. The hydrogen peroxide reacts with the iodide ions to produce iodine, which immediately reacts with the thiosulfate ions. When all the thiosulfate ions have been used up, the next iodine will produce an intense blue-black colour with the starch. At this point the clock is stopped. The experiment is then repeated, either at a different temperature or with the concentration of one of the reactants changed. The rate of reaction is approximately proportional to 1/time.

Interpretation of kinetic data

Concentration–time graphs

The most usual graphs have the amount or the concentration of a reactant on the y-axis and time on the x-axis. The slope of this type of graph at any point is the rate of the reaction at that point.

Worked example 1

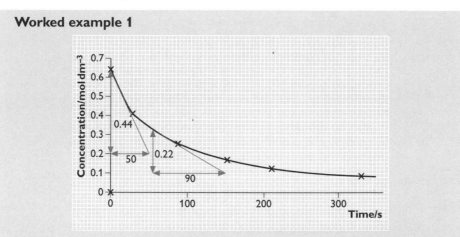

The slope at the initial concentration is $(0.64 - 0.20)/50 = 0.0088$ mol dm^{-3} s^{-1}.

The slope at 0.32 mol dm^{-3} = $(0.32 - 0.10)/(150 - 60) = 0.0024$ mol dm^{-3} s^{-1}.

This value is close to $\frac{1}{4}$ that of the initial rate. As the rate decreased by a factor of 4 when the concentration halved, the reaction is second-order.

Worked example 2

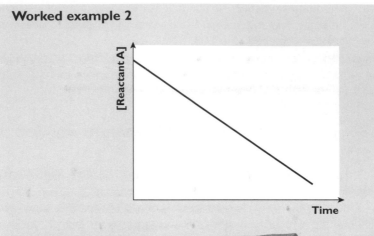

The slope is constant, therefore the rate is constant, and the reaction is zero-order with respect to A.

Worked example 3

The half-life is the time taken for the *concentration of a reactant to halve.* For a first-order reaction, the half-life is *constant.* Its value can be determined from a graph of concentration against time.

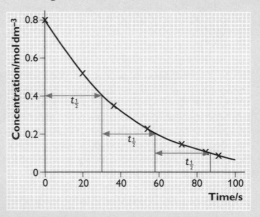

The time taken for the concentration to halve from 0.8 to 0.4 mol dm^{-3} is 30 s.

The time taken for the concentration to halve from 0.4 to 0.2 mol dm^{-3} is 28 s.

The time taken for the concentration to halve from 0.2 to 0.1 mol dm^{-3} is 29 s.

The half-lives are *constant* to within experimental error and so the reaction is first-order. The average half-life is (30 + 28 + 29)/3 = 29 s.

Note: the second half-life is the time for the concentration to fall from 0.4 to 0.2 mol dm^{-3}. This is 28 s. It is not the time for the concentration to fall from 0.8 to 0.2, i.e. 58 s. This is a common error.

content guidance

Worked example 4

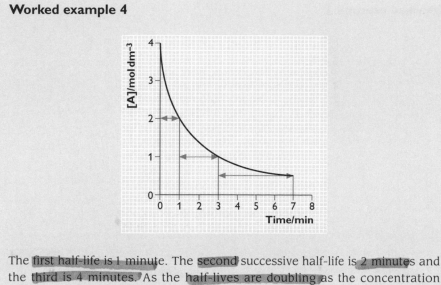

The first half-life is 1 minute. The second successive half-life is 2 minutes and the third is 4 minutes. As the half-lives are doubling as the concentration halves, the reaction is second-order with respect to A.

Rate–concentration graphs

The most usual graphs have the rate of reaction on the *y*-axis and concentration of a reactant or the concentration of a reactant squared on the *x*-axis.

If a straight line is obtained, the rate of reaction is proportional to the quantity plotted on the *x*-axis.

Note: for gases, the partial pressure (see page 33) rather than the concentration might be plotted. The partial pressure of a gas is proportional to its concentration.

Worked example 1

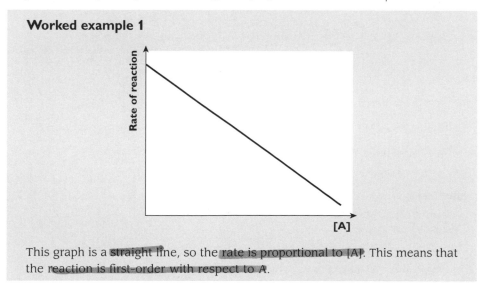

This graph is a straight line, so the rate is proportional to [A]. This means that the reaction is first-order with respect to A.

Worked example 2

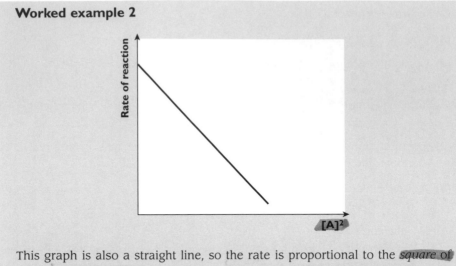

This graph is also a straight line, so the rate is proportional to the *square of* the concentration of A. This means that the reaction is second-order with respect to A.

Deducing the order of reaction from initial rate data

To do this, you need data from at least three experiments. For example, data from the reaction below could be used:

xA + yB $\longrightarrow$ products

Experiment	[A]/mol dm^{-3}	[B]/mol dm^{-3}	Initial rate/mol dm^{-3} s^{-1}
1	0.1	0.1	p
2	0.2	0.1	q
3	0.2	0.2	r

Comparing experiments 1 and 2 shows that [A] has doubled but [B] has stayed the same.

If the rate is unaltered ($q = p$), the order with respect to A is 0.

If the rate doubles ($q = 2p$), the order with respect to A is 1.

If the rate quadruples ($q = 4p$), the order with respect to A is 2.

The order with respect to B can be determined in a similar way, by analysing experiments 2 and 3.

Worked example
In the table above, suppose p = 0.0024, q = 0.0096 and r = 0.0096.

(1) Determine the partial orders of A and B.

(2) Determine the total order.

(3) Write the rate equation.

(4) Calculate the value of the rate constant.

Answer

(1) From experiments 1 and 2: when [A] doubles, the rate quadruples (0.0096 = 4 × 0.0024). Therefore, the order with respect to A is 2.

(2) From experiments 2 and 3: when [B] doubles, the rate is unaltered. Therefore, the order with respect to B is zero. The total order is 2 + 0 = 2.

(3) The rate equation is:
$$rate = k[A]^2$$

(4) The rate constant is:
$$k = rate/[A]^2 = 0.0024 \ mol \ dm^{-3} \ s^{-1}/(0.1 \ mol \ dm^{-3})^2 = 0.24 \ mol^{-1} \ dm^3 \ s^{-1}$$

Iodine–propanone reaction

In acid solution iodine reacts with propanone to form iodopropanone and hydrogen iodide.

$$CH_3COCH_3 + I_2 \xrightarrow{\ H^+(aq)\ } CH_3COCH_2I + HI$$

This reaction can be monitored by the following method.

- Place 25 cm³ of iodine solution of known concentration in a flask and add 25 cm³ of sulfuric acid solution (an excess) and 50 cm³ of water. Add 5 cm³ of pure propanone from a burette, starting the clock as you do so.
- At noted intervals of time remove 10 cm³ of the reaction mixture and quench it by adding it to 25 cm³ of aqueous sodium hydrogencarbonate. (This reacts with the acid catalyst, stopping the reaction.)
- Quickly titrate the remaining iodine against standard sodium thiosulfate solution, using starch as an indicator.
- Repeat by removing samples at regular intervals.

The rate equation is of the form:
$$rate = k[propanone]^x[H^+]^y[I_2]^z$$

Because a large excess of propanone was taken, the concentration of propanone, written as [propanone], remains approximately constant. Even though acid is produced in the reaction, the original amount of acid catalyst was large and so [H⁺] remains effectively constant. This means that the rate equation becomes:

$$rate = k'[I_2]^z \ where \ k' = k[propanone]^x[H^+]^y$$

The volume of sodium thiosulfate is proportional to the amount of iodine and hence to the iodine concentration. So a graph of volume of sodium thiosulfate against time will have the same shape as the graph of [I₂] against time.

The result of this experiment is shown in the graph below.

As this graph is a straight line, the slope and hence the rate is constant and so the reaction is zero order with respect to iodine.

A conclusion that can be drawn from this is that iodine must enter the mechanism *after* the rate-determining step.

Note: the order with respect to propanone can also be discovered. If the experiment is repeated with twice as much propanone, then [propanone] will be doubled. The slope of the graph of $[I_2]$ against time is twice that of the former graph. This shows that k' has doubled, so the reaction must be first order with respect to propanone.

Rate constant

Calculation of the rate constant
You must first find the order of reaction.

From rate data
If you know the rate of reaction between substances A and B at given concentrations and the order of reaction with respect to A and B, the rate constant can be found using the equation:

k = rate of reaction$/[A]^m[B]^n$

where m and n are the known orders.

From the half-life
If you have plotted a graph of concentration against time and found that the half-life is constant, the reaction is first-order and the rate constant is given by the expression:

$k = \ln 2/t$

Rate constant units

Total order	Units of k
Zero	mol dm^{-3} s^{-1}
First	s^{-1}
Second	mol^{-1} dm^3 s^{-1}

Effect of temperature on rate

An increase in temperature *always* increases the rate of a reaction because the rate constant increases with temperature. The mathematical relationship is given by the Arrhenius equation.

$$k = Ae^{-E_a/RT} \text{ or } \ln k = \ln A - E_a/RT$$

where A is a constant, E_a is the activation energy, R is the gas constant and T is the temperature in kelvin, where 25°C = (25 + 273) K. $°C \rightarrow K \ (x+273)$

An increase in temperature causes E_a/RT to become smaller. Thus E_a/RT becomes less negative and hence $\ln k$ and rate increase.

Note: you do not need to learn this equation. You will be given the equation if you need it.

A change in temperature does *not* alter the value of the activation energy. However, the Arrhenius equation also shows the relationship between activation energy and rate constant. A larger value of E_a gives a more negative exponential power and hence a smaller value of k. A *higher* activation energy results in a *slower* reaction.

A catalyst provides an alternative route with a lower activation energy. The *lower* E_{cat} results in a *larger k* and hence a *faster* reaction.

Calculation of activation energy

The correct way to do this is to measure the value of the rate constant, k, at different temperatures. A graph is then plotted of $\ln k$ against 1/temperature, where the temperature must be in kelvin. The slope of this graph = $-E_a/R$. The gas constant, R, has a value of 8.31 J K^{-1} mol^{-1}. *gradient*

Note: you can plot $\log_{10} k$ against 1/temperature, in which case the slope = $-E_a/2.3R$.

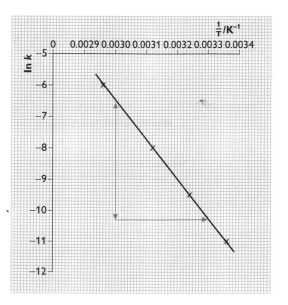

The slope (gradient) $= \dfrac{-10.3 - (-6.5)}{0.00330 - 0.00300} = \dfrac{-3.8}{0.00030} = -1.27 \times 10^4$ K

$E_a = -$ slope $\times R = -(-1.27 \times 10^4 \times 8.31) = +1.05 \times 10^5$ J mol^{-1} $= +105$ kJ mol^{-1}

Note: be careful about signs. The slope is calculated as $(y_2 - y_1)/(x_2 - x_1)$, where y_2 is the second value on the y-axis and x_2 the corresponding value on the x-axis.

Reaction mechanisms

Partial orders and mechanisms

- Reactions often take place in two or more steps. The slowest step is called the **rate-determining step**.
- If the order with respect to a reactant is zero, it must appear in the mechanism *after* the rate-determining step.
- If a reactant has a partial order of 2, *two* molecules of it appear in the mechanism *up to and including* the rate-determining step.

The correctness of a mechanism can be tested by comparing it with the experimentally determined partial orders.

Worked example

The following reaction was found to be first-order with respect to both A and B:

$2A + B \longrightarrow C + D$

Which of the following mechanisms is consistent with the data?

Mechanism I

Step 1:	A + B $\longrightarrow$ intermediate	Fast
Step 2:	intermediate + A $\longrightarrow$ C + D	Slow R·D·S

Mechanism II:

Step 1:	A + B $\longrightarrow$ intermediate	Slow R·D·S
Step 2:	intermediate + A $\longrightarrow$ C + D	Fast

Answer

In mechanism I, A appears twice and B once up to and including the rate-determining step. Therefore, the reaction would be second-order in A and first-order in B. This does not agree with the data.

In mechanism II, A and B both appear only once up to and including the rate-determining step. Therefore, the reaction would be first-order in both A and B, which is in agreement with the data.

Nucleophilic substitution reactions of halogenoalkanes

There are two possible mechanisms when a nucleophile, such as an OH$^-$ ion, reacts with a halogenoalkane.

One mechanism has an initial rate-determining step involving both the nucleophile and the halogenoalkane and proceeds via a transition state.

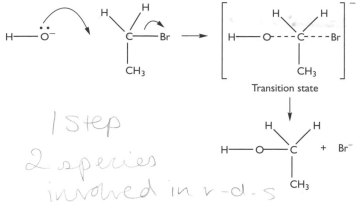

Transition state

1 step

2 species

involved in r-d.s

One piece of evidence for this mechanism is if the rate equation is of the form:

rate = *k*[halogenoalkane][nucleophile]

This mechanism is called an S_N2 mechanism, where S stands for substitution, N for nucleophilic and 2 for the number of species involved in the rate-determining step.

Another mechanism has an initial rate-determining step that only involves the halogenoalkane and not the nucleophile. This proceeds via an intermediate.

Step 1: the carbon–halogen bond breaks, a carbocation is formed and a halide ion is released. This is the slower rate-determining step.

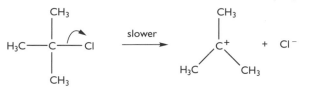

2 steps

Step 2: the carbocation is attacked by the nucleophile in a faster reaction.

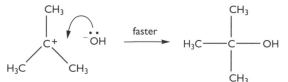

1 species involved in rate determining step

Evidence for this is the rate equation of the form:

rate = *k*[halogenoalkane]

This is called an S_N1 mechanism.

Further evidence can sometimes be obtained from the optical activity of the product (see page 65).

Primary halogenoalkanes react almost totally by an S_N2 mechanism, but tertiary halogenoalkanes react almost entirely by the S_N1 route. The rate of S_N2 decreases from primary to tertiary as the nucleophile is increasingly blocked by alkyl groups (steric effect). The rate of S_N1 increases from primary to tertiary because the intermediate planar carbocation is stabilised by 'electron-pushing' groups, such as the CH_3 group.

The reaction-profile diagrams for these two different types of reaction are shown below.

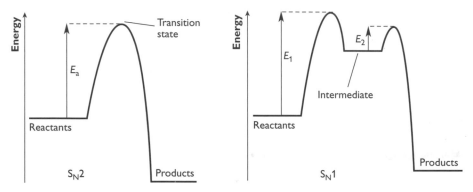

Note that as E_1 is greater than E_2, step 1 is rate-determining.

How far? Entropy

It is often assumed that only the value of the enthalpy change determines whether a reaction takes place. There are many examples of processes that are endothermic yet react to completion and others that do not react even though a calculated value of ΔH is highly negative. The direction of change is determined by the thermodynamics of the process.

There are three laws of thermodynamics.
- The first is that energy cannot be created or destroyed — the law of conservation of energy.
- The second is that, in any spontaneous process, the *total* entropy change is positive, which can be written as $\Delta S_{total} > 0$.
- The third law is that the entropy of any perfectly crystalline substance at a temperature of absolute zero (0 K or –273.16°C) is zero.

The entropy, S, of a substance is determined by how random the arrangement and energy of the molecules of that substance are. Thus:
- The entropy of any substance increases with temperature.
- A gaseous substance has a higher entropy than the less randomly arranged liquid form of that substance which, in turn, has a higher entropy than the highly ordered solid.

- S(water vapour) $> S$(liquid water) $> S$(ice)
- A complex substance has a higher entropy than a simple substance in the same physical state. Examples are that the entropy of $CaCO_3(s)$ is greater than that of CaO(s) and the entropy of $NH_3(g)$ is greater than that of $H_2(g)$.

Tables of entropies of substances are quoted under standard conditions of a stated temperature, usually 298 K, and 1 atmosphere pressure.

Entropy changes in a reaction

When any chemical reaction takes place, the entropy of the products will be different from the entropy of the reactants. The difference is called the entropy change of the *system*, ΔS_{system}, and is given by the equation:

$$\Delta S_{system} = \Sigma S_{products} - \Sigma S_{reactants}$$

> **Worked example**
>
> Calculate ΔS_{system} for the reaction of magnesium with oxygen to form magnesium oxide:
>
> $$2Mg(s) + O_2(g) \longrightarrow 2MgO(s) \qquad \Delta H = -1203 \text{ kJ mol}^{-1}$$
>
> Standard entropies (in J K^{-1} mol^{-1}) at 298 K are Mg(s) 32.7, $O_2(g)$ 205 and MgO(s) 26.9.
>
> **Answer**
> $$\Delta S_{system} = 2 \times S(MgO) - [2 \times S(Mg) + S(O_2)] = 2 \times 26.9 - [2 \times 32.7 + 205]$$
> $$= -216.6 \text{ J K}^{-1} \text{ mol}^{-1}$$

This is a negative number, but the reaction happens rapidly and completely when the magnesium is ignited. This shows that the entropy change of the *system* is not the only factor that determines whether a reaction is thermodynamically spontaneous or feasible.

What has not yet been considered is that the heat given off in this exothermic reaction increases the disorder of the surroundings.

The *total* entropy change is made up of two parts: the entropy change of the chemicals — the entropy change of the system — and the entropy change of the surroundings, ΔS_{surr}. This is shown by the equation:

$$\Delta S_{total} = \Delta S_{system} + \Delta S_{surr}$$

ΔS_{system} can be evaluated from data tables as in the worked example above. ΔS_{surr} is calculated from the relationship:

$$\Delta S_{surr} = -\Delta H/T, \text{ where the temperature } T \text{ must be in kelvin.}$$

So:

$$\Delta S_{total} = \Delta S_{system} - \Delta H/T$$

Do not forget the minus sign in this expression. Exothermic reactions have a negative ΔH and so a positive ΔS_{surr}.

Note: entropy values are normally measured in J K^{-1} mol^{-1}, whereas enthalpy changes are measured in kJ mol^{-1}.

Worked example

Calculate the value of ΔS_{surr} for the reaction between magnesium and oxygen. Use this value and the value of ΔS_{system} from the worked example above to calculate ΔS_{total} and comment on whether the reaction is thermodynamically spontaneous (feasible).

Answer

$\Delta H = -1203$ kJ mol^{-1}

$\Delta S_{surr} = -\Delta H / T = -(-1203 \times 1000/298) = +4036.9$ J K^{-1} mol^{-1}

$\Delta S_{total} = \Delta S_{system} + \Delta S_{surr} = -216.6 + (+4036.9) = +3820$ J K^{-1} mol^{-1}

The large positive value for ΔS_{total} shows that this reaction is thermodynamically spontaneous and will go to completion. This is in spite of the negative ΔS_{system}.

Summary

- Exothermic reactions have a positive ΔS_{surr} and are always spontaneous if ΔS_{system} is positive and may be spontaneous even if ΔS_{system} is negative.
- Endothermic reactions have a negative ΔS_{surr} and so are spontaneous only if ΔS_{system} is positive and is bigger in magnitude than ΔS_{surr}.

Effect of temperature on ΔS_{total}

If the temperature is increased, the *magnitude* of ΔS_{surr} decreases. This means that a positive ΔS_{surr} gets less positive and a negative ΔS_{surr} gets less negative.

Consider the reactions:

$$2Mg(s) + O_2(g) \longrightarrow 2MgO(s) \qquad \Delta H = -1203 \text{ kJ mol}^{-1}$$

$$2CH_3COOH(l) + (NH_4)_2CO_3(s) \longrightarrow 2CH_3COONH_4(s) + CO_2(g) + H_2O(l)$$

$$\Delta H = +60.5 \text{ kJ mol}^{-1}$$

For the combustion of magnesium:

ΔS_{surr} at 298 K $= -(-1\,203\,000)/298 = +4037$ J K^{-1} mol^{-1}

ΔS_{surr} at 373 K $= -(-1\,203\,000)/373 = +3225$ J K^{-1} mol^{-1}

The decrease in the positive value of ΔS_{surr} for the combustion of magnesium means that the value of ΔS_{total} also decreases (gets less positive).

For the reaction of ethanoic acid with ammonium carbonate:

ΔS_{surr} at 298 K $= -(+60\,500)/298 = -203$ J K^{-1} mol^{-1}

ΔS_{surr} at 373 K $= -(+60\,500)/373 = -162$ J K^{-1} mol^{-1}

The decrease in the negative value of ΔS_{surr} for the reaction of ethanoic acid with ammonium carbonate means that the value of ΔS_{total} increases (gets more positive).

Summary
- An increase in temperature for an exothermic reaction causes a decrease in ΔS_{total} and hence makes the reaction less thermodynamically likely.
- An increase in temperature for an endothermic reaction causes an increase in ΔS_{total} and hence makes the reaction more thermodynamically likely.

This explains the effect of a change in temperature on the value of the equilibrium constant and hence the position of equilibrium (see page 42).

Kinetic stability

The value of ΔS_{total} predicts whether the reaction is thermodynamically spontaneous, that is, whether it is feasible. Not all reactions that are thermodynamically feasible at a given temperature take place. If the activation energy is too high, too few molecules possess that energy on collision and so the reaction does not take place at that temperature. This is where the reactants are said to be kinetically stable or inert. Magnesium and oxygen at 298 K are thermodynamically unstable relative to magnesium oxide, but the reaction is slow unless they are heated, when the magnesium catches fire. The reaction is said to be thermodynamically feasible but kinetically inert.

The activation energy for the endothermic reaction between ethanoic acid and ammonium carbonate is low and ΔS_{total} is positive, so the reaction takes place rapidly and completely at room temperature. The reactants are said to be thermodynamically and kinetically unstable.

Thermodynamics of dissolving

When an ionic solid dissolves in water, the entropy of the solute increases as it changes from an ordered solid to a more random solution. If the ionic solid is anhydrous, it becomes hydrated on dissolving. Several water molecules, usually six, use the lone pair of electrons on the oxygen atom to bond exothermically with the cation. This makes the hydrogen atoms much more δ+ than in isolated water molecules and causes other water molecules to be joined on in a second sphere. Even more water molecules hydrogen bond to the second sphere, causing the water to become more ordered than in the pure state. The extent to which this happens depends on the charge density of the cation. Small, highly charged ions such as Al^{3+} or Mg^{2+} cause a considerable ordering of the water molecules, which outweighs the increase in randomness of the dissolved solute.

As a guide the following may be assumed:
- For singly positive cations, such as group 1, NH_4^+ and Ag^+, the ordering is not very large and is considerably outweighed by the increase in disorder of the solute. Thus ΔS_{system} for dissolving is positive.

- For doubly and triply charged cations, such as groups 2 and 3, and many transition metal ions, the ordering is significant and can outweigh the increase in randomness or disorder of the solute. Thus ΔS_{system} for dissolving is sometimes negative.

Whether an ionic solid dissolves in water or not is determined by the sign of ΔS_{total}. If this is positive, the solid dissolves. If negative, it is either insoluble or only slightly soluble.

The value of ΔS_{total} depends on the values of ΔS_{system} and ΔS_{surr} where $\Delta S_{surr} = -\Delta H/T$.

$$\Delta S_{total} = \Delta S_{system} - \Delta H/T$$

Calculation of ΔS_{system}

The absolute entropies of aqueous ions cannot be directly measured. Some data books give a value on a scale where the entropy, S, of $H^+(aq) = 0$. The calculation of ΔS_{system} is then done in the usual way.

Note: the justification for this is beyond A-level, but the results using these values are correct.

Worked example

Calculate the entropy change of the system for 1 mol of magnesium sulfate dissolving in water using the entropy values given in the table.

Species	Entropy value/J K^{-1} mol^{-1}
$MgSO_4(s)$	+91.6
$Mg^{2+}(aq)$	−138
$SO_4^{2-}(aq)$	+ 20.1

Answer

ΔS_{system} = S of $Mg^{2+}(aq)$ + S of $SO_4^{2-}(aq)$ − S of $MgSO_4(s)$
= −138 + (+20.1) − (+91.6) = −210 J K^{-1} mol^{-1}

Evaluation of $\Delta H_{solution}$

Some data books list values of $\Delta H_{solution}$. They can also be calculated using a Hess's Law cycle involving lattice energy and hydration energies.

Definitions

Enthalpy of solution, $\Delta H_{solution}$

This is the heat change when 1 mole of substance is dissolved in *excess* water.

For sodium chloride, this is represented by the equation:

$NaCl(s) + aq \longrightarrow Na^+(aq) + Cl^-(aq)$ $\Delta H_{solution} = +4$ kJ mol^{-1}

Lattice energy, ΔH_{latt} or LE

This is the energy change when 1 mole of ionic solid is formed from its gaseous ions infinitely far apart. It is always *exothermic*.

Using calcium hydroxide as an example, it is the heat change for the reaction:

$$Ca^{2+}(g) + 2OH^-(g) \longrightarrow Ca(OH)_2(s)$$

Hydration enthalpy, ΔH_{hyd}

This is the heat change, at constant pressure, when 1 mole of gaseous ions is dissolved in excess water to form an infinitely dilute solution.

Examples are:

$Na^+(g) + aq \longrightarrow Na^+(aq)$ $\Delta H_{hyd} = -406$ kJ mol^{-1}

$Cl^-(g) + aq \longrightarrow Cl^-(aq)$ $\Delta H_{hyd} = -364$ kJ mol^{-1}

Hydration enthalpy is always exothermic, because of the forces of attraction involved:

- Cations are bonded by the lone pair of electrons on the $\delta-$ oxygen atoms in water.
- Anions are surrounded by the $\delta+$ hydrogen atoms in water.

The Hess's law cycle for dissolving an ionic solid is:

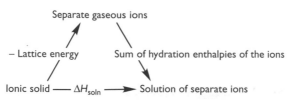

Thus:

$$\Delta H_{solution} = -(\text{lattice energy}) + (\text{sum of the hydration energies of the ions})$$

Worked example

Draw a Hess's law cycle for dissolving magnesium chloride in water and use it and the data in the table below to calculate the enthalpy of solution of magnesium chloride.

Enthalpy change	Value/kJ mol^{-1}
Lattice energy of MgCl$_2$(s)	-2493
Hydration enthalpy of Mg^{2+}(g)	-1920
Hydration enthalpy of Cl$^-$(g)	-364

Answer

The Hess's law cycle is:

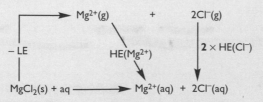

$$\Delta H_{soln}(MgCl_2) = -\Delta H_{latt}(MgCl_2) + (\text{sum of } \Delta H_{hyd} \text{ of Mg}^{2+} \text{ and 2Cl}^-)$$

$$= -(-2493) + [-1920 + (2 \times -364)] = -155 \text{ kJ mol}^{-1}$$

Note: enthalpy of solution refers to substances, whereas hydration enthalpy refers to ions.

Factors determining lattice energy
Ionic charge
The greater the charge, the stronger is the force between the positive and the negative ions. Therefore, the lattice energy is larger (more exothermic). For example, the lattice energy of magnesium fluoride (MgF_2), which contains the Mg^{2+} ion, is much larger (more exothermic) than that of sodium fluoride (NaF), which contains the Na^+ ion.

Sum of the ionic radii
The smaller the sum of the ionic radii, the stronger is the force between the positive and the negative ions. Therefore, the lattice energy is larger. For example:
- The lattice energy of sodium fluoride is larger than that of sodium chloride because the fluoride ion has a smaller radius than the chloride ion.
- The lattice energies of the hydroxides or sulfates of group 2 decrease down the group as the cations get larger.

Factors determining hydration enthalpy
Ionic charge
The greater the charge, the stronger is the ion/dipole force of attraction between the ion and the water molecules. Therefore, the hydration enthalpy is larger (more exothermic). For example, the hydration enthalpy of the magnesium ion, Mg^{2+}, is larger than that of the sodium ion, Na^+.

Ionic radius
The smaller the radius of the ion, the stronger is the force of attraction between the ion and the water molecules. Therefore, the hydration enthalpy is larger. For example:
- The hydration enthalpy of the fluoride ion, F^-, is larger than that of the chloride ion, Cl^-.
- The hydration enthalpy of a group 2 cation decreases as the group is descended ($Be^{2+} > Mg^{2+} > Ca^{2+} > Sr^{2+} > Ba^{2+}$).

Solubility of ionic solids

A solute is soluble in a given solvent if ΔS_{total} for dissolving is positive.

Group 1 substances
ΔS_{system} is positive. Even if $\Delta H_{solution}$ is positive (endothermic), $-\Delta H/T$ is smaller in magnitude than ΔS_{system}, so ΔS_{total} is positive and all group 1 solids are water-soluble.

Group 2 sulfates
ΔS_{system} is negative (see page 30) and gets less negative down the group as the ions get bigger in radius. The change in ΔS_{system} going from magnesium sulfate to barium sulfate is 108 J K^{-1} mol^{-1} less negative. The enthalpies of solution change from -91 kJ mol^{-1} for $MgSO_4$ to $+10$ kJ mol^{-1} for $BaSO_4$, so ΔS_{surr} changes from $+305$ to -64 J K^{-1} mol^{-1}. This is a decrease of 369 J K^{-1} mol^{-1}, and is a much larger decrease

than the change in ΔS_{system}. This causes ΔS_{total} to change from a positive value (soluble) for magnesium sulfate to a negative value (insoluble) for barium sulfate.

For magnesium sulfate:

$\Delta S_{total} = \Delta S_{system} - \Delta H/T = -210 - (-91\ 000/298) = +95\ J\ K^{-1}\ mol^{-1}$

For barium sulfate:

$\Delta S_{total} = \Delta S_{system} - \Delta H/T = -102 - (+19\ 000/298) = -166\ J\ K^{-1}\ mol^{-1}$

Thus the solubility of the sulfates *decreases* down the group.

Group 2 hydroxides

ΔS_{system} is negative and gets less negative down the group as the ions get bigger in radius. $\Delta H_{solution}$ goes from being slightly positive for magnesium hydroxide to negative (exothermic) for barium hydroxide. Both of these changes increase the solubility.

For magnesium hydroxide:

$\Delta S_{total} = \Delta S_{system} - \Delta H/T = -223 - (+3000/298) = -233\ J\ K^{-1}\ mol^{-1}$

For barium hydroxide:

$\Delta S_{total} = \Delta S_{system} - \Delta H/T = -112 - (-52\ 000/298) = +62\ J\ K^{-1}\ mol^{-1}$

Thus the solubility of the hydroxides *increases* down the group.

Equilibria

Definitions

Dynamic equilibrium

When a mixture reaches equilibrium, the reactions do not stop. The rate of the forward reaction equals the rate of the reverse reaction and there is no further change in the concentrations of any of the substances involved.

Concentration

The **concentration** of a substance is defined as the number of moles in 1 dm^3 of solution. The symbol [A] means the concentration of substance A. The units are mol dm^{-3}.

Partial pressure

Partial pressure refers to gases. It is defined as the pressure that the gas would exert if it were alone in the vessel at the same temperature. It is calculated using the formula:

partial pressure = mole fraction × total pressure

where mole fraction = $\dfrac{\text{number of moles of that gas}}{\text{total number of moles of gas}}$

Air contains 21% oxygen by moles, so the mole fraction of oxygen is 0.21. The partial pressure of oxygen in air at 2 atm = 0.21 × 2 = 0.42 atm.

Equilibrium constant, K

The equilibrium constant is a measure of the extent to which a reversible reaction takes place at a given temperature. A large value for the equilibrium constant means that a reaction is almost complete. A value greater than 1 means that the reaction proceeds further from left to right than from right to left.

The equilibrium constant can be measured:
- in terms of concentrations (K_c)
- for gases, in terms of partial pressures (K_p)

Expression for K_c

For a **homogeneous** equilibrium, the value of K_c is determined from the *equilibrium* concentrations of all the reactants and products. (Homogeneous means that all the reactants and products are in the same phase, for instance all gases or all in solution.)

For the reaction:

$$x A + y B \rightleftharpoons m C + n D$$

where x, y, m and n are the stoichiometric amounts in the equation, the value of K_c at $T\,°C$ is given by:

$$K_c = \frac{[C]_{eq}{}^m [D]_{eq}{}^n}{[A]_{eq}{}^x [B]_{eq}{}^y}$$

- $[C]_{eq}{}^m$ is the concentration of C at equilibrium raised to the power m.
- The temperature should *always* be quoted, because the value of an equilibrium constant varies with temperature.
- The units of K_c can be determined from the expression by cancelling out the units for the separate concentration terms.
- The expression for K_c is only valid when *equilibrium concentrations* are used.
- The quotient $[C]^m[D]^n/[A]^x[B]^y$ equals K_c only when the system is at equilibrium.
- The quotient is also called the concentration term.

The relationship for a particular reaction can be shown by measuring the concentrations of the species at equilibrium. The experiment is repeated using several different starting concentrations.

For example, in the reaction between hydrogen and iodine:

$$H_2(g) + I_2(g) \rightleftharpoons 2HI(g)$$

Known amounts of hydrogen and iodine are heated to a certain temperature and allowed to reach equilibrium. The system is then analysed to find the amounts of each substance present.

The experiment is repeated using different ratios of hydrogen and iodine and heated to the same temperature. Whatever the starting amounts, the relationship:

$$\frac{[HI]_{eq}{}^2}{[H_2]_{eq}{}^2 [I_2]_{eq}{}^2}$$

is always equal to the same number — the equilibrium constant, K_c.

Worked example 1

State the expression, and the units, of K_c for the reaction:

$$Br_2(aq) + 2Fe^{2+}(aq) \rightleftharpoons 2Br^-(aq) + 2Fe^{3+}(aq)$$

Answer

$$K_c = \frac{[Br^-]_{eq}{}^2[Fe^{3+}]_{eq}{}^2}{[Br_2]_{eq}[Fe^{2+}]_{eq}{}^2}$$

The units are:

$$\frac{(mol\ dm^{-3})^2 \times (mol\ dm^{-3})]^2}{(mol\ dm^{-3}) \times (mol\ dm^{-3})^2} = \frac{(mol\ dm^{-3})^4}{(mol\ dm^{-3})^3} = mol\ dm^{-3}$$

Worked example 2

Consider the reaction:

$$2SO_2(g) + O_2(g) \rightleftharpoons 2SO_3(g) \qquad K_c = 1.7 \times 10^6\ mol^{-1}\ dm^3\ at\ 700\ K$$

A steel vessel, volume 2 dm^3, contains 0.2 mol SO_3, 0.04 mol SO_2 and 0.01 mol O_2. Show whether or not this mixture is at equilibrium. If not, indicate which direction the system will move in order to achieve equilibrium at 700 K.

Answer

$$K_c = \frac{[SO_3]_{eq}{}^2}{[SO_2]_{eq}{}^2[O_2]_{eq}} = 1.7 \times 10^6\ mol^{-1}\ dm^3$$

$$[SO_3] = \frac{0.2}{2} = 0.1\ mol\ dm^{-3}$$

$$[SO_2] = \frac{0.04}{2} = 0.02\ mol\ dm^{-3}$$

$$[O_2] = \frac{0.01}{2} = 0.005\ mol\ dm^{-3}$$

$$quotient = \frac{[SO_3]^2}{[SO_2]^2[O_2]} = \frac{0.1^2}{0.02^2 \times 0.005} = 5.0 \times 10^3\ mol^{-1}\ dm^3$$

This value does not equal K_c. Therefore, the system is *not* at equilibrium.

The quotient (concentration term) is *less than* K_c. Therefore, its value must increase to achieve equilibrium. This means that the reaction will move to the right, increasing the concentration of SO_3 and decreasing the concentrations of SO_2 and O_2, until the value of the quotient equals 1.7×10^6. The system is then in equilibrium.

Calculation of K_c

In K_c calculations, you are always given the chemical equation plus some data. These data could be concentrations at equilibrium, in which case all you have to do is to put the values into the expression for K_c. However, you are usually given the *initial* amounts and either the amount of one substance at equilibrium or the percentage of a reactant converted. You must use *equilibrium* concentrations, not initial concentrations, in your calculation.

The calculation should be carried out in five steps.
- **Step 1** — use the chemical equation to write the expression for K_c.
- **Step 2** — construct a table and write in the initial number of moles of each substance.
- **Step 3** — write in the table the change in moles of each substance and hence work out the number of moles at equilibrium.
- **Step 4** — divide the equilibrium number of moles by the volume to get the concentration in mol dm^{-3}.
- **Step 5** — substitute the *equilibrium* concentrations into the expression for K_c and work out its value. At the same time, work out the units of K_c and include them in your answer.

Worked example

Phosphorus pentachloride (PCl_5) decomposes according to the equation:

$$PCl_5(g) \rightleftharpoons PCl_3(g) + Cl_2(g)$$

1.0 mol of phosphorus pentachloride was added to a 20 dm^3 flask and heated to 180°C. When equilibrium had been established, it was found that 32% of the phosphorus pentachloride had decomposed. Calculate the value of K_c at 180°C.

Answer

Step 1: $K_c = \dfrac{[PCl_3]_{eq}[Cl_2]_{eq}}{[PCl_5]_{eq}}$

		PCl$_5$	**PCl$_3$**	**Cl$_2$**	**Units**
Step 2	Initial amount	1.0	0	0	mol
Step 3	Change	−0.32	+0.32	+0.32	mol
	Moles at equilibrium	1.0 − 0.32 = 0.68	0.32	0.32	mol
Step 4	Concentration at equilibrium	$\dfrac{0.68}{20} = 0.034$	$\dfrac{0.32}{20} = 0.016$	$\dfrac{0.32}{20} = 0.016$	mol dm^{-3}

Step 5:
$$K_c = \frac{[PCl_3]_{eq}[Cl_2]_{eq}}{[PCl_5]_{eq}} = \frac{0.016 \text{ mol dm}^{-3} \times 0.016 \text{ mol dm}^{-3}}{0.034 \text{ mol dm}^{-3}} = 0.0075 \text{ mol dm}^{-3}$$

Note that:
- as 32% of the PCl_5 reacted and the initial amount was 1 mol, 0.32 × 1 = 0.32 mol reacted. Therefore, the change in moles of PCl_5 was −0.32, so 0.68 mol (1.0 − 0.32) of PCl_5 remained in the equilibrium mixture.
- the molar ratio was 1:1:1. Therefore, if 0.32 mol of PCl_5 reacted, 0.32 mol of PCl_3 and 0.32 mol of Cl_2 were formed.

The calculation is more complicated if the stoichiometric ratio in the equation is not 1:1. Consider the equilibrium:

$$CH_4(g) + 2H_2O(g) \rightleftharpoons CO_2(g) + 4H_2(g)$$

For each mole of methane that reacts, 2 mol of steam are required and 1 mol of carbon dioxide and 4 mol of hydrogen are produced.

Worked example

Chlorine can be produced from hydrogen chloride by the following reaction:

$$4HCl(g) + O_2(g) \rightleftharpoons 2Cl_2(g) + 2H_2O(g)$$

0.40 mol of hydrogen chloride and 0.10 mol of oxygen were placed in a vessel of volume 4.0 dm^3 and allowed to reach equilibrium at 400°C. At equilibrium, only 0.040 mol of hydrogen chloride was present. Calculate the value of K_c.

Answer

$$K_c = \frac{[Cl_2]_{eq}^2[H_2O]_{eq}^2}{[HCl]_{eq}^4[O_2]_{eq}}$$

	4HCl	**O$_2$**	**2Cl$_2$**	**2H$_2$O**
Initial moles	0.40	0.10	0	0
Change	−0.36	$-\frac{1}{4}(0.36) = 0.09$	$+\frac{1}{2}(0.36) = +0.18$	$+\frac{1}{2}(0.36) = +0.18$
Moles at equilibrium	0.40 − 0.36 = 0.04	0.10 − 0.09 = 0.01	0.18	0.18
Concentration at equilibrium	$\frac{0.04}{4.0} = 0.010$	$\frac{0.01}{4.0} = 0.0025$	$\frac{0.18}{4.0} = 0.045$	$\frac{0.18}{4.0} = 0.045$

The moles of hydrogen chloride decreased from 0.40 to 0.040, which is a decrease of 0.36 mol.

$$K_c = \frac{[Cl_2]_{eq}^2[H_2O]_{eq}^2}{[HCl]_{eq}^4[O_2]_{eq}} = \frac{(0.045 \text{ mol dm}^{-3})^2 \times (0.045 \text{ mol dm}^{-3})^2}{(0.010 \text{ mol dm}^{-3})^4 \times 0.0025 \text{ mol dm}^{-3}} = 1.6 \text{ mol}^{-1} \text{ dm}^3$$

Note that:

- There were 0.040 moles of HCl at equilibrium and the initial amount was 0.40 mol. Therefore, the amount of HCl that reacted was 0.40 − 0.040 = 0.36 mol. So the change in moles of HCl was −0.36.
- The change in moles of O$_2$ was one quarter of the change in moles of HCl, because the ratio of O$_2$ to HCl in the equation was 1:4.
- The change in moles of Cl$_2$ (and of H$_2$O) was half the change in moles of HCl, because the ratio of moles of Cl$_2$ to HCl in the equation was 2:4.

Expression for K_p

K_p is the equilibrium constant expressed in terms of partial pressures.

For a homogeneous gas phase reaction:

$$x A(g) + y B(g) \rightleftharpoons m C(g) + n D(g)$$

The expression for K_p is:

$$K_p = \frac{p(C)^m \, p(D)^n}{p(A)^x \, p(B)^y}$$

where $p(A)$ is the partial pressure of substance A *at equilibrium*.

The units of K_p can be worked out from the expression for K_p by cancelling the units for the individual partial pressure terms.

Remember that the partial pressure of a gas A in a mixture of gases A, B and C is given by:

$$p(A) = \frac{\text{moles of A}}{\text{moles of A} + \text{moles of B} + \text{moles of C}} \times \text{total pressure}$$

The sum of the partial pressures of all the gases in the system equals the total pressure:

$$p(A) + p(B) + p(C) = \text{total pressure}$$

Worked example

For the equilibrium below, state the expression for K_p and give its units.

$$N_2(g) + 3H_2(g) \rightleftharpoons 2NH_3(g)$$

Answer

$$K_p = \frac{p(NH_3)^2}{p(N_2)\, p(H_2)^3}$$

The units are:

$$\frac{\text{atm}^2}{\text{atm} \times \text{atm}^3} = \frac{\text{atm}^2}{\text{atm}^4} = \text{atm}^{-2}$$

Calculation of K_p

In K_p calculations, you are always given the chemical equation plus some data, usually the initial amounts and either the amount of one substance at equilibrium or the percentage of a reactant converted. You will also be given the total pressure. You *must* use equilibrium amounts, *not* initial amounts, in your calculation.

The calculation should be carried out in six steps.

- **Step 1** — use the chemical equation to write the expression for K_p.
- **Step 2** — construct a table and write in the initial number of moles of each substance.
- **Step 3** — write in the table the change in moles of each substance and hence work out the number of moles at equilibrium. Add up the moles to find the *total* number of gas moles.

- **Step 4** — divide the number of moles of each substance at equilibrium by the total number of moles to obtain the **mole fraction**.
- **Step 5** — multiply the mole fraction by the *total* pressure to obtain the partial pressure of each gas.
- **Step 6** — substitute the partial pressures into the expression for K_p and work out its value. At the same time, work out the units of K_p and include them in your answer.

Tip For a decomposition equilibrium, such as $PCl_5 \rightleftharpoons PCl_3 + Cl_2$, some questions simply state that a percentage of the reactant has reacted. If so, assume that you started with 1 mol and work from that value.

Worked example

Consider the equilibrium:

$$2SO_2(g) + O_2(g) \rightleftharpoons 2SO_3(g)$$

When 1.6 mol of sulfur dioxide was mixed with 1.2 mol of oxygen and allowed to reach equilibrium at 450°C, 90% of the sulfur dioxide reacted. The total pressure in the vessel was 7.0 atm. Calculate the value of K_p at this temperature.

Answer

Step 1: $K_p = \dfrac{p(SO_3)^2}{p(SO_2)^2 \times p(O_2)}$

		$2SO_2$	**O_2**	**$2SO_3$**	**Total**
Step 2	Initial moles	1.6	1.2	0	–
Step 3	Change	-90% of $1.6 = -1.44$	$-\frac{1}{2} \times 1.44 = -0.72$	$+1.44$	–
	Moles at equilibrium	$1.6 - 1.44 = 0.16$	$1.2 - 0.72 = 0.48$	1.44	2.08
Step 4	Mole fraction	$\dfrac{0.16}{2.08} = 0.0769$	$\dfrac{0.48}{2.08} = 0.231$	$\dfrac{1.44}{2.08} = 0.692$	–
Step 5	Partial pressure/atm	$0.0769 \times 7 = 0.538$	$0.231 \times 7 = 1.62$	$0.692 \times 7 = 4.84$	–

Step 6:

$$K_p = \frac{p(SO_3)^2}{p(SO_2)^2 \times p(O_2)} = \frac{(4.84 \text{ atm})^2}{(0.538 \text{ atm})^2 \times 1.62 \text{ atm}} = 50 \text{ atm}^{-1}$$

Note that:

- 90% of the SO_2 reacted, so 0.90×1.6 mol $= 1.44$ mol reacted $=$ moles of SO_3 produced
- the ratio of O_2 to SO_2 is 1:2, so the moles of O_2 that reacted $= \frac{1}{2} \times 1.44 = 0.72$
- the total number of moles is $0.16 + 0.48 + 1.44 = 2.08$

Heterogeneous equilibria

In a heterogeneous equilibrium, the reactants and products are in two different phases, for example a gas phase and a solid phase. Solid substances do *not* appear in the expression for the equilibrium constant, because solids do not have a partial pressure.

Consider the reaction:

$$2Fe(s) + 3H_2O(g) \rightleftharpoons Fe_2O_3(s) + 3H_2(g)$$

The expression for K_p is:

$$K_p = \frac{p(H_2)^3}{p(H_2O)^3} \text{ and it has no units}$$

Consider the reaction:

$$CaO(s) + CO_2(g) \rightleftharpoons CaCO_3(s)$$

The expression for K_p is:

$$K_p = \frac{1}{p(CO_2)} \text{ and it has units of atm}^{-1}$$

Distribution of a solute between two immiscible solvents

If a solute such as iodine is shaken between water and a hydrocarbon solvent, an equilibrium is reached where:

$$\frac{[I_2] \text{ in the hydrocarbon solvent}}{[I_2] \text{ in water}} = \text{a constant } K_c$$

This constant is sometimes called the distribution constant or partition constant. Care must be taken to use concentrations not moles, if the volume of the water is different from that of the hydrocarbon solvent.

Relationship between the total entropy change and the equilibrium constant

$$\Delta S_{total} = R \ln K$$

where R is the gas constant and equals 8.31 J K^{-1} mol^{-1}.

Note that:

$$\Delta S_{total} = \Delta S_{system} + \Delta S_{surr} = \Delta S_{system} - \Delta H/T$$

This means that, as a change in temperature causes a change in ΔS_{total}, the value of the equilibrium constant, K, will change as well.

An increase in temperature always causes a *decrease* in the magnitude of ΔS_{surr} (makes it less positive or less negative — see page 28).

For an exothermic reaction (ΔH negative, hence ΔS_{surr} positive):
An increase in temperature causes the ΔS_{surr} to become less positive, so ΔS_{total} becomes smaller. This, in turn, makes the value of K smaller.

For an endothermic reaction (ΔH positive, so ΔS_{surr} negative):
An increase in temperature causes the ΔS_{surr} to become less negative, so ΔS_{total} becomes larger. This, in turn, makes the value of K larger.

Extent of reaction

How far a reaction goes is determined by the magnitude of the equilibrium constant. The value of the entropy change and the equilibrium constant give an estimate of how complete a reaction is at a particular temperature.

The relationship between ΔS_{total}, K and extent of reaction is shown in the table below.

ΔS_{total}/J K^{-1} mol^{-1}	K	Extent of reaction
> +150	> 10^8	Reaction almost complete
Between +60 and +150	Between 10^3 and 10^8	Reaction favours products
Between +60 and −60	Between 10^3 and 10^{-3}	Both reactants and products in significant quantities
Between −60 and −150	Between 10^{-3} and 10^{-8}	Reaction favours reactants
More negative than −150	< 10^{-8}	Reaction not noticeable

Applications of rates and equilibrium

Factors affecting K and the position of equilibrium

It is important to realise that the position of equilibrium shifts so that K and the quotient (concentration or partial pressure term) are equal again.

Catalyst

A catalyst:
• has *no* effect on the value of an equilibrium constant

- does not alter the position of equilibrium
- speeds up the forward and reverse reactions equally, so that equilibrium is reached more rapidly

Temperature

Exothermic reactions

- An increase in temperature of an exothermic reaction decreases the value of the equilibrium constant (see page 41). This means that the quotient (or concentration term) is now not equal to the equilibrium constant, so the system is no longer at equilibrium.
- The position of equilibrium then shifts to the left, causing the quotient to decrease in value until it equals the new lower value of the equilibrium constant.
- This causes a lowering of the equilibrium yield.

Endothermic reactions

- An increase in temperature of an endothermic reaction causes an increase in the value of the equilibrium constant (see page 41). This means that the quotient is now smaller than the new value of the equilibrium constant and the system is no longer in equilibrium.
- The position of equilibrium then shifts to the right, causing the quotient to increase in value until it equals the new higher value of the equilibrium constant.
- This causes an increase in the equilibrium yield of product.

It is vital to realise that the logic of the effect of temperature is:

Temperature alters — this causes the value of K to change — this causes the position of equilibrium to alter.

It is *not*:
Temperature alters — position of equilibrium shifts (according to Le Chatelier) — K alters.

Note: a decrease in temperature has the opposite effect.

Consider the reaction:

$$A \rightleftharpoons B \qquad \Delta H = -123 \text{ kJ mol}^{-1}$$

$$K_c = \frac{[B]_{eq}}{[A]_{eq}}$$

If the temperature is increased, the value of K_c decreases as the reaction is exothermic. Thus it becomes smaller than the quotient [B]/[A]. To restore equilibrium, the quotient must also become smaller. Therefore, some B reacts to form A, until the quotient [B]/[A] once again equals K_c and the system is back in equilibrium.

Pressure

- A change in pressure has *no* effect on the value of the equilibrium constant.

- For reactions with a different number of gas moles on each side of the equation, an increase in pressure will cause the quotient to alter in value.
- The system is now no longer in equilibrium and so the system will react, altering the value of the quotient, until it equals the unaltered value of the equilibrium constant.
- The result is that an increase in pressure will move the position of equilibrium to the side with fewer gas moles or a decrease in pressure will move the position of equilibrium to the side with more gas moles.

Consider the reaction:

$$A(g) + 2B(g) \longrightarrow C(g)$$

An increase in pressure does *not* alter the value of the equilibrium constant. However, it causes the quotient $[C]/[A][B]^2$ to decrease as all the concentrations increase, but the bottom line increases more than the top line. The system then reacts so that the quotient increases, until it equals the unaltered value of K_c. The result is that the position of equilibrium shifts to the right (the side with fewer gas moles), thus increasing the equilibrium yield of C.

Justification of conditions in industrial processes

The Haber process

$$N_2(g) + 3H_2(g) \rightleftharpoons 2NH_3(g)$$

$$\Delta H = -92.4 \text{ kJ mol}^{-1} \qquad K_p = 1 \times 10^{-5} \text{ atm}^{-2} \text{ at } 400°C$$

The conditions are:
- temperature — 400°C
- pressure — 200 atm
- catalyst — iron

If a lower temperature were used, the equilibrium yield of ammonia would be increased. This is because a decrease in temperature causes the equilibrium constant to increase in value, as the reaction is exothermic. Thus it is now bigger than the quotient $p(NH_3)^2/p(N_2)p(H_2)^3$. To restore equilibrium, the quotient must get larger. Therefore some nitrogen and hydrogen react to form ammonia, increasing the value of the quotient until it equals the new larger value of the equilibrium constant. Thus the position of equilibrium would shift to the right, increasing the equilibrium yield of ammonia.

However, a lower temperature means a lower rate, so a compromise temperature is used as well as the iron catalyst. Thus the reaction can take place quickly at a temperature that results in a fast reaction and an acceptable yield of ammonia.

At 1 atm and 400°C, the percentage of ammonia in the equilibrium mixture is less than 0.1%. This is because of the very small value of the equilibrium constant, K_p.

To counter this, a pressure of 200 atm is used. An increase in pressure causes the quotient to get smaller as there are fewer gas moles on the right. The quotient is now less than the unaltered value of K_p and the system is therefore not at equilibrium. To regain equilibrium, nitrogen and hydrogen react to form ammonia, until the quotient once again equals the unaltered K_p. Use of such a high pressure is expensive in terms of the capital and running costs of the plant, but it is essential as otherwise the yield of ammonia would be uneconomic.

Even under these conditions, the yield is less than 20%. So the mixture is cooled, the ammonia liquefies and the unreacted gaseous nitrogen and hydrogen are mixed with more nitrogen and hydrogen and are recycled through the catalyst. In this way almost all the hydrogen is finally converted into ammonia, resulting in a high atom economy.

Worked example

The critical reaction in the manufacture of hydrogen is between methane and steam.

$$CH_4(g) + H_2O(g) \rightleftharpoons CO(g) + 3H_2(g) \qquad \Delta H = +206 \text{ kJ mol}^{-1}$$

The conditions which give about a 90% conversion are:
- temperature: 750°C
- pressure: 2 atm
- catalyst: nickel

Comment on the choice of conditions.

Answer

The reaction is endothermic, so an increase in temperature will result in an increase in the value of K and hence an increase in the equilibrium yield of hydrogen. However a higher temperature would cause engineering problems and so is not used. Even at 750°C the reaction is slow and so a catalyst is used.

A low pressure would not cause a change in the value of K but would cause the quotient to decrease in value, as there are more gas moles on the right. This in turn would result in a higher yield of hydrogen, as the reaction would move to the right increasing the value of the quotient until it equals the unaltered value of K. However, there must be sufficient excess pressure to drive the gases through the plant, and so a lower pressure than 2 atm would not be possible.

Note: in both these examples, an increase in pressure would not alter the rate. This is typical of heterogeneous gas reactions with a solid catalyst. The rate, at a given temperature, is dependent on the number of active sites on the surface of the catalyst and not on the pressure.

Acid–base equilibria

Early theories

Acids were first identified by their sour taste. It was then found that acids had a common effect on certain vegetable extracts such as litmus or red cabbage. The next step was the thought that all acids contained oxygen. In fact the name oxygen means 'sharp-maker' in Greek. Later the Dutch chemist Arrhenius suggested that acids produced hydrogen ions in aqueous solution.

$$HA(aq) + H_2O(l) \longrightarrow H_3O^+(aq) + A^-(aq)$$

This definition is still used today but only applies to aqueous solutions. Brønsted and Lowry expanded this theory to include a number of similar reactions.

Brønsted–Lowry theory of acidity

Acids

An acid is a proton (H^+ ion) donor; it gives an H^+ ion to a base.

Hydrogen chloride is an acid, giving an H^+ ion to a base such as OH^-.

$$HCl + OH^- \longrightarrow H_2O + Cl^-$$

An acid also protonates water, which acts as a base.

$$HCl + H_2O \longrightarrow H_3O^+ + Cl^-$$

Strong acids protonate water completely, as in the example above.

Weak acids, such as ethanoic acid, protonate water reversibly.

$$CH_3COOH + H_2O \rightleftharpoons H_3O^+ + CH_3COO^-$$

Bases

A base accepts H^+ ions. To do this, it must have a lone pair of electrons.

Ammonia is a base, accepting an H^+ ion from an acid such as hydrogen chloride.

$$NH_3 + HCl \longrightarrow NH_4^+ + Cl^-$$

Ammonia is a weak base. It is protonated *reversibly* by water, which acts as an acid.

$$NH_3 + H_2O \rightleftharpoons NH_4^+ + OH^-$$

Conjugate pairs

In a reaction an acid loses an H^+ ion. The species resulting from the loss of an H^+ is called the **conjugate base** of that acid.

Acid	Conjugate base
HCl	Cl^-
H_2SO_4	HSO_4^-
H_2O	OH^-
CH_3COOH	CH_3COO^-
NH_4^+	NH_3

The species resulting from a base gaining an H^+ ion is called the **conjugate acid** of that base.

Base	Conjugate acid
OH^-	H_2O
H_2O	H_3O^+
NH_3	NH_4^+
$C_2H_5NH_2$	$C_2H_5NH_3^+$
HSO_4^-	H_2SO_4

The equation for an acid–base reaction has an acid on the left with its conjugate base on the right, and a base on the left with its conjugate acid on the right. In this type of reaction there are two acid–base conjugate pairs that you must be able to identify.

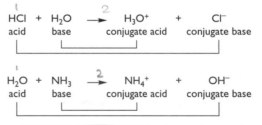

- An acid is linked to its conjugate base by the loss of an H^+ ion.
- A base is linked to its conjugate acid by the gain of an H^+ ion.

Note: water can act either as a base or as an acid.

Worked example

Concentrated nitric acid reacts with concentrated sulfuric acid according to the equation:

$$HNO_3 + H_2SO_4 \longrightarrow H_2NO_3^+ + HSO_4^-$$

Identify the acid–base conjugate pairs in this reaction.

Answer

First pair: acid, H_2SO_4; its conjugate base, HSO_4^-

Second pair: base, HNO_3; its conjugate acid, $H_2NO_3^+$

pH

Definition of pH

The pH of a solution is defined as the negative logarithm to the base 10 of the hydrogen ion concentration in mol dm^{-3}.

$$pH = -\log[H^+]$$

Thus, if $[H^+] = 1.23 \times 10^{-2}$ mol dm^{-3}, the pH $= -\log[H^+] = -\log(1.23 \times 10^{-2}) = 1.91$.

Tip Always give pH values to two decimal places. If you are given the pH, you can calculate $[H^+]$ using the expression $[H^+] = 10^{-pH}$. Thus, if pH = 2.44, $[H^+] = 10^{-2.44} = 0.00363$ mol dm^{-3}. You must know how to work this out on your calculator. Remember that $[H^+]$ will normally be less than 1 mol dm^{-3}.

Definition of pOH

The pOH of a solution is defined as the negative logarithm to the base 10 of the hydroxide ion concentration.

$$pOH = -\log[OH^-]$$

K_w and pK_w

Water partially ionises.

$$H_2O \rightleftharpoons H^+ + OH^-$$

$$K_c = \frac{[H^+][OH^-]}{[H_2O]}$$

The concentration of H_2O is so large that it is effectively constant. Therefore, its value can be incorporated into the expression for the equilibrium constant. Thus, the dissociation constant for water, K_w, is defined as:

$$K_w = [H^+][OH^-] = 1.0 \times 10^{-14} \text{ at } 25°C$$

$$pK_w = -\log K_w = 14$$

$$pK_w = -\log([H^+][OH^-]) = pH + pOH$$

At 25°C, pH + pOH = 14

pH of pure water, acidic solutions and alkaline solutions

- Pure water is neutral.
  - A *neutral* solution is defined as one in which $[H^+] = [OH^-]$.
  - Therefore, $[H^+] = [OH^-] = \sqrt{K_w} = \sqrt{1.0 \times 10^{-14}} = 1.0 \times 10^{-7}$ mol dm^{-3} at 25°C.
  - Therefore, the pH of a neutral solution at 25°C = $-\log(1.0 \times 10^{-7}) = 7.00$.
- An *acidic* solution is defined as one in which $[H^+] > [OH^-]$.
  - Therefore, at 25°C, $[H^+] > 1.0 \times 10^{-7}$, which means that pH < 7.
- An *alkaline* solution is defined as one where $[H^+] < [OH^-]$.
  - Therefore, at 25°C, $[H^+] < 1.0 \times 10^{-7}$, resulting in a pH > 7.
  - If a solution has $[H^+] = 3.0 \times 10^{-9}$ mol dm^{-3}, the pH = $-\log(3.0 \times 10^{-9}) = 8.52$ and it is alkaline.
- If the temperature is not 25°C:
  - The ionisation of water is endothermic, so an increase in temperature will result in an increase in the value of K_w. At 100°C, the value of $K_w = 5.13 \times 10^{-13}$ mol^2 dm^{-6}.
  - A neutral solution at 100°C will have $[H^+] = \sqrt{K_w} = \sqrt{5.13 \times 10^{-13}} = 7.16 \times 10^{-7}$ mol dm^{-3}, and so a pH = $-\log(7.16 \times 10^{-7}) = 6.15$.

Calculation of pH of an alkaline solution

The pH of an alkaline solution can be calculated from $[OH^-]$. For example:

$[OH^-] = 2.2 \times 10^{-2}$ mol dm^{-3}

$pOH = -\log[OH^-] = -\log(2.2 \times 10^{-2}) = 1.66$

$pH + pOH = 14$

$pH = 14 - pOH = 14 - 1.66 = 12.34$

An alternative method is:

$[OH^-] = 2.2 \times 10^{-2}$ mol dm^{-3}

$[H^+][OH^-] = 1.0 \times 10^{-14}$

$[H^+] = \dfrac{1.0 \times 10^{-14}}{[OH^-]} = \dfrac{1.0 \times 10^{-14}}{2.2 \times 10^{-2}} = 4.55 \times 10^{-13}$

$pH = -\log[H^+] = -\log(4.55 \times 10^{-13}) = 12.34$

pH scale

- Neutral solutions: $[H^+] = [OH^-]$; pH = 7.00 (at 25°C)
- Acidic solutions: $[H^+] > [OH^-]$; pH < 7
- Alkaline solutions: $[H^+] < [OH^-]$; pH > 7

The pH scale runs from negative numbers to above 14. A small change in hydrogen ion or hydroxide ion concentration causes a large pH change in a solution with an initial pH close to 7. A pH change from 1 to 0 is the result of a change in hydrogen ion concentration from 0.1 to 1 mol dm⁻³ (a change of 0.9 mol dm⁻³). A pH change from 6 to 5 is the result of a change in hydrogen ion concentration from 0.000 001 to 0.000 01 mol dm⁻³, an increase of only 0.000 009 mol dm⁻³.

pH at 25°C	$[H^+]$/mol dm^{-3}	Acidity
Negative	> 1	Very strongly acidic
0 to 2	1 to 10^{-2}	Strongly acidic
2 to 5	10^{-2} to 10^{-5}	Weakly acidic
6	10^{-6}	Very weakly acidic
7	10^{-7}	Neutral
pH at 25°C	$[OH^-]$/mol dm^{-3}	Alkalinity
7	10^{-7}	Neutral
8	10^{-6}	Very weakly alkaline
9 to 12	10^{-5} to 10^{-2}	Weakly alkaline
12 to 14	10^{-2} to 1	Strongly alkaline
Above 14	> 1	Very strongly alkaline

Strong acids and bases

Strong acids

A strong acid, such as hydrochloric acid, is *totally* ionised in aqueous solution.

$HCl + aq \longrightarrow H^+(aq) + Cl^-(aq)$ or $HCl(aq) + H_2O(l) \longrightarrow H_3O^+(aq) + Cl^-(aq)$

Thus, $[H^+]$ = the initial concentration of the acid.
- Hydrochloric acid, hydrobromic acid, HBr, and hydroiodic acid, HI, are all strong acids.
- Nitric acid, HNO_3, is also a strong acid.
- Sulfuric acid, H_2SO_4, is strong from its *first* ionisation only.
   $H_2SO_4 + aq \longrightarrow H^+(aq) + HSO_4^-(aq)$
- The pH of an acid can be negative if its concentration is more than 1 mol dm⁻³. For instance, the pH of a 2.00 mol dm⁻³ solution of HCl = –log(2.00) = –0.30.

Worked example
Calculate the pH of the strong monobasic acid HCl of concentration 0.123 mol dm⁻³.

Answer
pH = –log$[H^+]$ = –log(0.123) = 0.91

Strong bases

Strong bases are totally ionised in solution. For example:

$$NaOH + aq \longrightarrow Na^+(aq) + OH^-(aq)$$

$$Ba(OH)_2 + aq \longrightarrow Ba^{2+}(aq) + 2OH^-(aq)$$

For strong bases with one OH^- ion per formula, the OH^- ion concentration is equal to the initial concentration of the base. For bases with two OH^- ions per formula, the OH^- ion concentration is twice the initial concentration of the base.

There are two ways of calculating the pH.

Worked example 1

Calculate the pH of a 2.00 mol dm^{-3} solution of sodium hydroxide.

Answer using method 1

$[OH^-]$ = 2.00 mol dm^{-3}

pOH = $-\log[OH^-]$ = $-\log(2.00)$ = -0.30

pH = 14 $-$ pOH = 14 $-$ (-0.30) = 14.30

Answer using method 2

$[OH^-]$ = 2.00 mol dm^{-3}

$$[H^+] = \frac{1.0 \times 10^{-14}}{[OH^-]} = \frac{1.0 \times 10^{-14}}{2.00} = 5.00 \times 10^{-15}$$

pH = $-\log[H^+]$ = $-\log(5.00 \times 10^{-15})$ = 14.30

Worked example 2

Calculate the pH of a 0.222 mol dm^{-3} solution of Ba(OH)$_2$.

Answer using method 1

$[OH^-]$ = 2 $\times$ 0.222 = 0.444 mol dm^{-3}

pOH = $-\log[OH^-]$ = $-\log(0.444)$ = 0.35

pH = 14 $-$ pOH = 14 $-$ 0.35 = 13.65

Answer using method 2

$[OH^-]$ = 2 $\times$ 0.222 = 0.444 mol dm^{-3}

$$[H^+] = \frac{1.0 \times 10^{-14}}{[OH^-]} = \frac{1.0 \times 10^{-14}}{0.444} = 2.25 \times 10^{-14} \text{ mol dm}^{-3}$$

pH = $-\log[H^+]$ = $-\log(2.25 \times 10^{-14})$ = 13.65

Weak acids

A weak acid is only *slightly* ionised. For dilute solutions of most weak acids, the extent of ionisation is less than 10%, so most of the acid is present as un-ionised molecules.

$$CH_3COOH + aq \rightleftharpoons H_3O^+(aq) + CH_3COO^-(aq)$$

Acid dissociation constant of a weak acid, K_a

The expression for the acid dissociation constant of a weak acid and questions depending on it are easier to follow if the formula of a weak acid is represented by HA.

HA ionises reversibly according to:

$$HA \rightleftharpoons H^+ + A^- \quad \text{or} \quad HA + H_2O \rightleftharpoons H_3O^+ + A^-$$

The acid dissociation constant, K_a, is given by:

$$K_a = \frac{[H^+][A^-]}{[HA]} \quad \text{or} \quad K_a = \frac{[H_3O^+][A^-]}{[HA]} \quad \text{units: mol dm}^{-3}$$

Do *not* include $[H_2O]$ in the expression for K_a.

The pK_a of a weak acid is defined as $-\log K_a$, so $K_a = 10^{-pK_a}$.

Tip You can use H^+, $H^+(aq)$ or H_3O^+ in the chemical equation and in the expression for K_a. As the acid is only slightly ionised, you can make the approximation:

[HA] = the initial concentration of the weak acid

The ionisation of 1 molecule of HA produces one H^+ ion and one A^- ion and so, ignoring the tiny amount of H^+ from the water:

$$[H^+] = [A^-]$$

The expression for K_a becomes:

$$K_a = \frac{[H^+]^2}{[HA]}$$

$$[H^+] = \sqrt{K_a[HA]} = \sqrt{K_a \times \text{initial concentration of weak acid}}$$

Therefore, the pH of a weak acid can be calculated from its concentration and its acid dissociation constant.

Worked example

Calculate the pH of a 0.123 mol dm^{-3} solution of ethanoic acid.

$K_a = 1.74 \times 10^{-5}$ mol dm^{-3}

Answer

$$CH_3COOH \rightleftharpoons H^+ + CH_3COO^-$$

$$K_a = \frac{[H^+][CH_3COO^-]}{[CH_3COOH]} = \frac{[H^+]^2}{CH_3COOH} = 1.74 \times 10^{-5} \text{ mol dm}^{-3}$$

$$[H^+] = \sqrt{K_a[CH_3COOH]} = \sqrt{1.74 \times 10^{-5} \times 0.123} = 0.00146 \text{ mol dm}^{-3}$$

$$pH = -\log[H^+] = -\log(0.00146) = 2.83$$

Tip The pK_a of the acid may be given instead of its K_a value. In this case, use $K_a = 10^{-pK_a}$ — for example:

pK_a for ethanoic acid = 4.76

K_a for ethanoic acid = $10^{-4.76} = 1.74 \times 10^{-5}$ mol dm^{-3}

Note: if you are given the mass of ethanoic acid and the volume of solution, you must first calculate the concentration of the solution in mol dm^{-3}.

$$\text{concentration} = \text{moles} \div \text{volume in dm}^3 = \frac{\text{mass of acid}}{\text{molar mass of acid}} \div \frac{\text{volume in cm}^3}{1000}$$

You must understand the two assumptions made here:

(1) [CH$_3$COOH] in solution = [acid] originally (ignoring the small amount ionised)
(2) [H$^+$] = [CH$_3$COO$^-$] (ignoring the small amount of H$^+$ produced by the ionisation of water)

Calculation of K_a from pH

This is a similar calculation to the previous one. If you are given the pH and the concentration of a solution, K_a can be calculated.

> **Worked example**
> Calculate the acid dissociation constant for a weak acid, HA, given that a 0.106 mol dm^{-3} solution has a pH of 3.21.
>
> *Answer*
>
> $$K_a = \frac{[H^+][A^-]}{[HA]}$$
>
> [H$^+$] = 10^{-pH} = 10$^{-3.21}$ = 6.166 × 10^{-4} mol dm^{-3} = [A$^-$]
>
> [HA] = 0.106 − 0.0006166 = 0.1054 mol dm^{-3}
>
> $$K_a = \frac{(6.166 \times 10^{-4})^2}{0.1054} = 3.61 \times 10^{-6} \text{ mol dm}^{-3}$$
>
> As 6.166 × 10^{-4} moles of H$^+$ ions are produced from 0.106 mol of HA, the amount of HA at equilibrium equals (0.106 − 6.166 × 10^{-4}) = 0.1054 mol.
>
> If [HA] = 0.106 mol dm^{-3} had been used, the value of K_a would have been calculated as 3.59 × 10^{-6} mol dm^{-3} — a less accurate answer that would also have scored full marks.

Tip In all calculations involving just the weak acid, you may assume that [H$^+$] = [A$^-$]. This is not true with buffer solutions where the conjugate base is also present (see page 58).

pH of sulfuric acid

Sulfuric acid is a strong acid from its first ionisation, but a weak acid from its second.

$$H_2SO_4 + aq \longrightarrow H^+ + HSO_4^-$$

$$HSO_4^- + aq \rightleftharpoons H^+ + SO_4^{2-}$$

The second ionisation is driven to the left (suppressed) by the high concentration of

H^+ from the first ionisation. Hence, the pH of a 0.1 mol dm^{-3} solution of sulfuric acid is only just below 1, as 0.1 mol dm^{-3} of H^+ ions is produced from the first ionisation and hardly any from the second ionisation.

Effect of dilution on pH

When a strong acid is diluted by a factor of 10, its pH increases by 1 unit.

When a weak acid is diluted by a factor of 10, its pH increases by half a unit.

Worked example 1

Calculate the pH of a 0.123 mol dm^{-3} solution of HCl and a 0.0123 mol dm^{-3} solution.

Answer

For the 0.123 mol dm^{-3} solution: pH = $-\log 0.123 = 0.91$

For the 0.0123 mol dm^{-3} solution: pH = $-\log 0.123 = 1.91$

Worked example 2

The pH of a 0.123 mol dm^{-3} solution of ethanoic acid, $K_a = 1.74 \times 10^{-5}$ mol dm^{-3} is 2.83 (see page 51).

Calculate the pH of a 0.0123 mol dm^{-3} solution of ethanoic acid.

Answer

$[H^+]^2 = K_a \times 0.0123 = 1.74 \times 10^{-5} \times 0.0123$

$[H^+] = \sqrt{(1.74 \times 10^{-5} \times 0.0123)} = 4.63 \times 10^{-4}$

pH = $-\log 4.63 \times 10^{-4} = 3.33$, which is an increase of half a pH unit from the pH of the solution ten times more concentrated.

pH of various solutions

If the pH of strong and weak acids, strong and weak bases and salts, all at a concentration of 0.1 mol dm^{-3}, are measured the following results are obtained.

Type	Example	pH
Strong acid	Hydrochloric acid (HCl)	1
Weak acid	Ethanoic acid (CH_3COOH)	Approx. 3
Strong base	Sodium hydroxide (NaOH)	13
Weak base	Ammonia (NH_3)	Approx. 11
Salt of a strong acid and a strong base	Sodium chloride (NaCl)	7
Salt of a weak acid and a strong base	Sodium ethanoate (CH_3COONa)	Approx. 9
Salt of a strong acid and a weak base	Ammonium chloride (NH_4Cl)	Approx. 5

The rule of two

Note that:
- a weak acid has a pH about 2 units more (less acidic) than a strong acid
- a weak base has a pH about 2 units less (less alkaline) than a strong base
- the salt of a weak acid has a pH about 2 units more than the salt of a strong acid
- the salt of a weak base has a pH about 2 units less than the salt of a strong base

Acid–base titrations

There are several factors to consider.
- The usual method is to add a standard solution of base from a burette to a known volume of acid, in the presence of a suitable indicator. The addition of base is normally stopped when the 'end point' has been reached.
- The end point is reached when enough base has been added to react totally with the acid. The pH at the end point is 7 only when a strong base is titrated with a strong acid.
- The word 'neutralise' is often used. However, the end point is on the acidic side of neutral if a weak base, such as ammonia solution, is titrated with a strong acid and above 7 if a strong base is titrated with a weak acid, such as ethanoic acid.

Titration	pH at end point	Common indicators
Strong acid–strong base	7	Methyl orange, methyl red or phenolphthalein
Strong acid–weak base	5–6	Methyl orange or methyl red
Weak acid–strong base	8–9	Phenolphthalein

- Acid–base titrations can also be carried out by adding a standard solution of acid from a burette to a known volume of base.

Titration curves

These need to be drawn carefully. Follow these steps:
- Decide whether the acid and base are strong or weak.
- Draw the axes. The y-axis (vertical) should be labelled 'pH', with a linear scale, and the x-axis labelled 'volume of solution being added' (normally the base). Some questions ask for the titration curve when acid is added to a base, in which case the x-axis represents the volume of acid.
- Calculate the volume that has to be added to reach the end point. In most questions, the concentrations of acid and base are the same; therefore, the volume of base at the end point will be the same as the initial volume of acid.
- Estimate the pH:
  – at the start
  – at the end point

– of the vertical range of the curve
– after excess base has been added (final pH)
- Draw a smooth curve with the vertical section at the end point volume.

The table below shows pH values during typical titrations.

Type of titration	Starting pH	End point pH	Vertical range of pH	Final pH
Strong base added to strong acid	1	7	3–11	Just <13
Strong base added to weak acid	3	9	7–11	Just <13
Weak base added to strong acid	1	5	3–7	Just <11
Strong acid added to strong base	13	7	11–3	Just >1
Strong acid added to weak base	11	5	7–3	Just >1
Weak acid added to strong base	13	9	11–7	Just >3

The graphs below are examples of titration curves.

Strong acid–strong base

The left-hand graph shows the change in pH when 40 cm³ of base is added to 20 cm³ of acid, both having the same concentration. The right-hand graph is for the addition of acid to 20 cm³ of base. The end point is 20 cm³. For example, hydrochloric acid and sodium hydroxide:

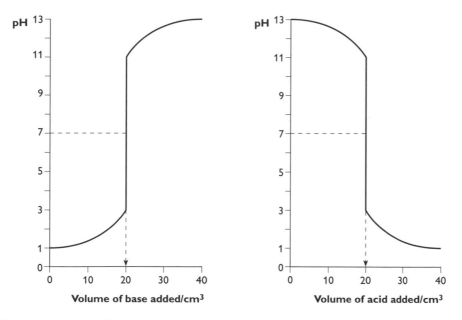

Strong base–weak acid

For example, sodium hydroxide added to ethanoic acid:

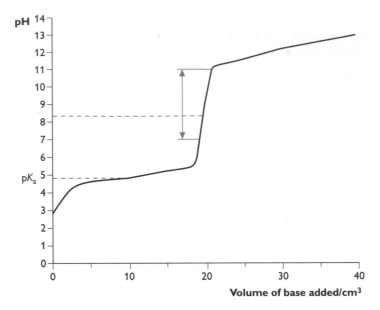

Strong acid–weak base

For example, hydrochloric acid added to ammonia:

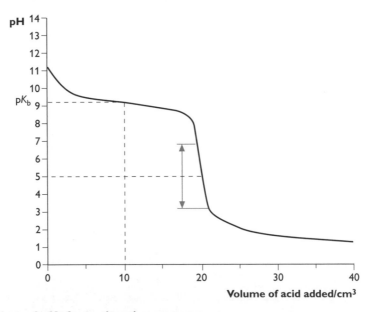

Calculation of pK_a from titration curves

When a strong base is added to a solution of a weak acid, the value of pK_a can be found by reading the pH at the point when *half* the acid has been neutralised by the alkali (halfway to the end-point volume).

At this point:

$$[HA]_{halfway} = [A^-]_{halfway}$$

$$K_a = \frac{[H^+][A^-]}{[HA]} = \frac{[H^+]_{halfway}[A^-]_{halfway}}{[HA]_{halfway}} = [H^+]_{halfway}$$

$$pK_a = (pH)_{halfway}$$

In the titration curve for a strong base being added to a weak acid, $pK_a = 4.5$.

Indicators

Indicators are weak acids in which the acid molecule, represented by the formula HInd, is a different colour from its conjugate base, Ind$^-$.

$$HInd \rightleftharpoons H^+ + Ind^-$$

An example is methyl orange. The acid, HInd, is red and its conjugate base, Ind$^-$, is yellow.

$$K_{ind} = \frac{[H^+][Ind^-]}{[HInd]}$$

At the end-point colour, the acid and its conjugate base are in equal proportions. Therefore:

$$[HInd] = [Ind^-]$$

$$K_{ind} = \frac{[H^+][Ind^-]}{[HInd]} = [H^+]_{end\ point}$$

$$pK_{ind} = (pH)_{end\ point}$$

Most indicators noticeably change colour over a range of ±1 pH unit. Methyl orange, $pK_{ind} = 3.7$, is red below a pH of 2.7, gradually changes to orange at pH 3.7 and then to yellow at pH 4.7.

Choice of indicator

The pH range over which an indicator changes colour must lie *completely within* the vertical part of the titration curve.

- For a strong acid–strong base titration, any indicator with pK_{ind} value from 3.5 to 10.5 has its range completely within the vertical part and is therefore suitable. Thus, methyl red ($pK_{ind} = 5.1$), methyl orange ($pK_{ind} = 3.7$) and phenolphthalein ($pK_{ind} = 9.3$) are all suitable.
- For a strong acid–weak base titration, the indicator must have a pK_{ind} value from 3.5 to 6.5 to be suitable. Thus, methyl red ($pK_{ind} = 5.1$) and methyl orange ($pK_{ind} = 3.7$) are both suitable.
- For a weak acid–strong base titration, the indicator must have a pK_{ind} value between pH 6.5 and 10.5. Phenolphthalein ($pK_{ind} = 9.3$) is suitable.

Buffer solutions

Definition

A buffer solution *resists* a change in pH when *small* amounts of acid or base are added. It is made up of a weak acid and its conjugate base, or a weak base and its conjugate acid. Examples are:

- ethanoic acid (CH_3COOH) plus sodium ethanoate ($CH_3COO^-Na^+$), which is an acid buffer with a pH less than 7
- ammonia solution (NH_3) plus ammonium chloride ($NH_4^+Cl^-$), which is an alkaline buffer with a pH greater than 7

A buffer solution is at its most efficient when:

- for an acidic buffer, the concentration of the weak acid equals the concentration of its conjugate base
- for an alkaline buffer, the concentration of the weak base equals the concentration of its conjugate acid

Tip Do not say that a buffer has constant pH, because the pH *does* alter slightly when small amounts of acid or base are added.

Calculation of pH of a buffer solution

Calculation of the pH of a buffer solution relies on the relationship between K_a and the concentrations of the weak acid and its salt.

$$K_a = \frac{[H^+][\text{salt of weak acid}]}{[\text{weak acid}]}$$

$$[H^+] = \frac{K_a[\text{weak acid}]}{[\text{salt of weak acid}]}$$

The pH is then calculated as $-\log[H^+]$.

Worked example 1

Calculate the pH of a buffer solution made by adding 6.15 g of sodium ethanoate, CH_3COONa (molar mass 82 g mol^{-1}), to 50.0 cm^3 of a 1.00 mol dm^{-3} solution of ethanoic acid ($K_a = 1.75 \times 10^{-5}$ mol dm^{-3}).

Answer

$$K_a = \frac{[H^+][CH_3COO^-]}{[CH_3COOH]}$$

or

$$[H^+] = \frac{K_a[CH_3COOH]}{[CH_3COO^-]}$$

$[CH_3COOH] = 1.00$ mol dm^{-3}

amount of sodium ethanoate $= \dfrac{6.15 \text{ g}}{82 \text{ g mol}^{-1}} = 0.075$ mol

$[CH_3COO^-] = [CH_3COONa] = \dfrac{0.075 \text{ mol}}{0.050 \text{ dm}^3} = 1.50 \text{ mol dm}^{-3}$

$[H^+] = 1.75 \times 10^{-5} \text{ mol dm}^{-3} \times \dfrac{1.00 \text{ mol dm}^{-3}}{1.50 \text{ mol dm}^{-3}} = 1.17 \times 10^{-5} \text{ mol dm}^{-3}$

$pH = -\log(1.17 \times 10^{-5}) = 4.93$

Worked example 2

Calculate the pH of a buffer solution made by mixing 100 cm³ of a 0.200 mol dm⁻³ solution of ethanoic acid, $K_a = 1.75 \times 10^{-5}$ mol dm⁻³, with 100 cm³ of a 0.300 mol dm⁻³ solution of sodium ethanoate.

Answer

$[CH_3COOH]$ in buffer $= \frac{1}{2} \times 0.200 = 0.100$ mol dm⁻³

$[CH_3COO^-]$ in buffer $= \frac{1}{2} \times 0.300 = 0.150$ mol dm⁻³

$[H^+] = K_a \dfrac{[\text{acid}]}{[\text{salt}]} = 1.75 \times 10^{-5} \times \dfrac{0.100}{0.150} = 1.17 \times 10^{-5}$ mol dm⁻³

$pH = -\log(1.17 \times 10^{-5}) = 4.93$

Note: the pH values of the buffer solutions in the two worked examples are the same. The concentration of acid and salt in example 2 is ten times less than the concentration in example 1. This illustrates the point that diluting a buffer does not alter its pH as the ratio of [acid]:[salt] remains the same.

A more difficult calculation is when a known volume of strong base solution is added to an *excess* of weak acid solution. All the base reacts with some of the weak acid. Thus a mixture of excess weak acid and its salt is formed, which is a buffer solution.

Worked example 3

Calculate the pH of a buffer solution made by adding 50 cm³ of 1.00 mol dm⁻³ solution of sodium hydroxide to 80 cm³ of 1.00 mol dm⁻³ solution of propanoic acid (CH_3CH_2COOH), $K_a = 1.35 \times 10^{-5}$ mol dm⁻³.

Answer

amount of sodium hydroxide added $= 1.00$ mol dm⁻³ $\times 0.050$ dm³

$= 0.050$ mol

$=$ amount of sodium propanoate produced

amount of propanoic acid initially $= 1.00$ mol dm⁻³ $\times 0.080$ dm³ $= 0.080$ mol

amount of propanoic acid *left* after reaction with sodium hydroxide $= 0.080 - 0.050 = 0.030$ mol

$[CH_3CH_2COOH] = \dfrac{0.030 \text{ mol}}{0.130 \text{ dm}^3} = 0.231$ mol dm⁻³

$[CH_3CH_2COO^-] = [CH_3CH_2COONa] = \dfrac{0.050 \text{ mol}}{0.130 \text{ dm}^3} = 0.385$ mol dm⁻³

$$K_a = \frac{[H^+][CH_3CH_2COO^-]}{[CH_3CH_2COOH]}$$

$$[H^+] = \frac{K_a[CH_3CH_2COOH]}{[CH_3CH_2COO^-]} = 1.35 \times 10^{-5} \text{ mol dm}^{-3} \times \frac{0.231 \text{ mol dm}^{-3}}{0.385 \text{ mol dm}^{-3}}$$

$$= 8.10 \times 10^{-6} \text{ mol dm}^{-3}$$

$$pH = -\log(8.10 \times 10^{-6}) = 5.09$$

Note that:

- propanoic acid and sodium hydroxide react in a 1:1 ratio to produce sodium propanoate and water
- the total volume of the solution is $50 + 80 = 130 \text{ cm}^3 = 0.130 \text{ dm}^3$

Summary of calculations of pH values of solutions of weak acids and buffers

Both are based on the expression for K_a:

$$K_a = \frac{[H^+][A^-]}{[HA]} \quad \text{or} \quad [H^+] = \frac{K_a[HA]}{[A^-]}$$

In *both* solutions: [HA] = [weak acid]

In a solution of a *weak acid*: $[H^+]$ = $[A^-]$

In a *buffer* solution: $[A^-]$ = [salt]

Tip The pH of a *weak acid* can be calculated using the formula $pH = \frac{1}{2}pK_a - \frac{1}{2}\log[HA]$.

The pH of a *buffer* solution can be calculated using the formula $pH = pK_a - \log([HA]/[A^-])$.

However, it is safer to start with the K_a expression as many mistakes are made by candidates in the exam when using the formulae above.

Mode of action of a buffer solution

Acid buffer

The ethanoic acid–sodium ethanoate buffer system is used as an example. The weak acid is *partially* ionised.

$$CH_3COOH \rightleftharpoons H^+ + CH_3COO^-$$

Its salt, which contains the conjugate base, is *totally* ionised.

$$CH_3COONa \longrightarrow CH_3COO^- + Na^+$$

The CH_3COO^- ions produced suppress the ionisation of the weak acid. Therefore, if the weak acid and its salt are present in similar molar amounts:

$[CH_3COOH]$ = [weak acid]

$[CH_3COO^-]$ = [salt of weak acid]

If a *small* quantity of H^+ ions is added, almost all react with the large reservoir of

CH_3COO^- ions from the salt:

$$H^+ + CH_3COO^- \rightleftharpoons CH_3COOH$$

The values of $[CH_3COO^-]$ and $[CH_3COOH]$ hardly alter, as they are both large *relative* to the *small* amount of H^+ added.

$$[H^+] = \frac{K_a[\text{acid}]}{[\text{salt}]} = \frac{K_a[CH_3COOH]}{[CH_3COO^-]}$$

If neither $[CH_3COOH]$ nor $[CH_3COO^-]$ alter significantly, the value of $[H^+]$ and hence the pH will also not alter significantly.

If a *small* amount of OH^- ions is added to the buffer solution, almost all of them react with the large reservoir of CH_3COOH molecules from the acid:

$$OH^- + CH_3COOH \longrightarrow H_2O + CH_3COO^-$$

Again, neither $[CH_3COOH]$ nor $[CH_3COO^-]$ alters significantly as both are large *relative* to the small amount of OH^- added, so the pH hardly alters.

The crucial points that must be made when explaining the mode of action of a buffer are that:

- the weak acid ionises *reversibly* (equation needed)
- the salt ionises *totally* (equation needed) and suppresses the ionisation of the acid, which results in reservoirs of both the weak acid and its conjugate base that are relatively large compared to the small amounts of H^+ or OH^- added
- almost all the added H^+ ions are removed by reaction with the conjugate base (equation needed)
- almost all the added OH^- ions are removed by reaction with the weak acid (equation needed)
- the ratio [weak acid]/[salt] does not change significantly because both quantities are large *relative* to the small amounts of H^+ or OH^- added

Further organic chemistry

Required AS chemistry

Nomenclature

You should refresh your knowledge of how organic compounds are named.

Carbon chain length	Stem name
One atom	Meth-
Two atoms	Eth-
Three atoms	Prop-
Four atoms	But-
Five atoms	Pent-

If the chain contains three or more carbon atoms, the position of a functional group is indicated by a number — for example, $CH_3CH_2CH=CH_2$ is but-1-ene; $CH_3CH(OH)CH_3$ is propan-2-ol.

If the carbon chain is *branched*, the position of the alkyl branch is indicated by a number. For example, $CH_3CH_2CH(CH_3)CH_2CH_2OH$ is 3-methylpentan-1-ol.

Bond enthalpy and polarity

The smaller the value of the bond enthalpy, the weaker is the bond. This means that the activation enthalpy of a reaction involving the breaking of that bond will be lower and the rate of reaction faster. For example, the C–I bond is weaker than the C–Cl bond, so iodoalkanes react faster than chloroalkanes.

The π-bond in alkenes is weaker than the σ-bond in alkanes, so alkenes are more reactive than alkanes.

The bond polarity determines the type of reaction. The carbon atom in halogenoalkanes is $\delta+$, so it is attacked by nucleophiles.

AS reactions, reagents and conditions

Alkanes

In the presence of light, alkanes react with chlorine — for example:

$$CH_4 + Cl_2 \longrightarrow CH_3Cl + HCl$$

The organic product is chloromethane.

• synoptic link: learn free-radical substitution mechanism

Alkenes

In the table below, propene is used as the example.

Reagent	Equation	Conditions	Organic product
Hydrogen	$CH_3CH=CH_2 + H_2 \longrightarrow$ $CH_3CH_2CH_3$	Heated nickel catalyst 150°C	Propane
Halogens e.g. Br_2	$CH_3CH=CH_2 + Br_2 \longrightarrow$ $CH_3CHBrCH_2Br$	Organic solvent	1,2-dibromopropane
Hydrogen halides e.g. HCl	$CH_3CH=CH_2 + HCl \longrightarrow$ $CH_3CHClCH_3$	Mix gases at room temperature	2-chloropropane
Oxidation e.g. $KMnO_4$	$CH_3CH=CH_2 + [O] + H_2O$ $\longrightarrow CH_3CH(OH)CH_2OH$	• Aqueous solution • with H_2SO_4	Propane-1,2-diol

Hydrogenation

Electrophilic Addition

Electrophilic Addition

Oxidation

Alcohols

In the table below ethanol is used as the example.

Reagent	Equation	Conditions	Organic product
Sodium	$CH_3CH_2OH + Na \longrightarrow$ $CH_3CH_2O^-Na^+ + \frac{1}{2}H_2$	Add sodium to ethanol	Sodium ethoxide
Phosphorus pentachloride	$CH_3CH_2OH + PCl_5 \longrightarrow$ $CH_3CH_2Cl + POCl_3 + HCl_{(g)}$	Dry at room temperature	Chloroethane
Hydrogen bromide	$CH_3CH_2OH + HBr \longrightarrow$ $CH_3CH_2Br + H_2O$	HBr is made by adding 50% sulfuric acid to solid potassium bromide	Bromoethane
Oxidation e.g. with acidified potassium dichromate	$CH_3CH_2OH + [O] \longrightarrow$ $CH_3CHO + H_2O$	Heat and distill out the aldehyde	Ethanal
	$CH_3CH_2OH + 2[O] \longrightarrow$ $CH_3COOH + H_2O$	Heat under reflux	Ethanoic acid

Halogenoalkanes

In the table below, 1-bromoethane is used as the example.

Reagent	Equation	Conditions	Organic product
Aqueous alkali e.g. NaOH(aq)	$CH_3CH_2Br + NaOH \longrightarrow$ $CH_3CH_2OH + NaBr$	Heat under reflux	Ethanol
Ethanolic alkali e.g. KOH	$CH_3CH_2Br + KOH \longrightarrow$ $CH_2{=}CH_2 + KBr + H_2O$	Heat under reflux in a solution of ethanol	Ethene
Water in the presence of silver nitrate solution	$CH_3CH_2Br + H_2O \longrightarrow$ $CH_3CH_2OH + H^+ + Br^-$ Then $Ag^+ + Br^- \longrightarrow AgBr$	Add to silver nitrate solution and observe ppt of AgBr	Ethanol
Ammonia	$CH_3CH_2Br + 2NH_3 \longrightarrow$ $CH_3CH_2NH_2 + NH_4Br$	Use ammonia dissolved in ethanol	Ethylamine

Isomerism

Isomers have the same molecular formula, but the atoms are arranged differently within the molecule.

Structural isomerism

- **Carbon chain** — the isomers have different carbon chain lengths. For instance, butane ($CH_3CH_2CH_2CH_3$) and methylpropane ($CH_3CH(CH_3)CH_3$) have the same molecular formula, C_4H_{10}.
- **Positional** — the same functional group is in a different position in the isomers, for example propan-1-ol ($CH_3CH_2CH_2OH$) and propan-2-ol ($CH_3CH(OH)CH_3$).
- **Functional group** — the isomers are members of different homologous series. Examples are:
  - propanoic acid (CH_3CH_2COOH) and methylethanoate (CH_3COOCH_3)
  - ethanol (C_2H_5OH) and methoxymethane (CH_3OCH_3)

Tip If you are asked to draw a *full* structural formula, you must draw each individual atom and bond.

Geometric isomerism

Geometric (*cis–trans*) isomerism is a form of **stereoisomerism**. In an alkene, geometric isomerism is the result of restricted rotation about a carbon–carbon double bond, provided that the two groups on each atom of the C=C group are different from each other.

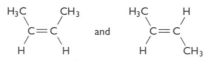

The π-overlap in a double bond is above and below the plane of the molecule. Therefore, it is not possible to rotate about the double bond without breaking the π-bond, which only happens at high temperatures. Hence, the *cis* and *trans* isomers are different.

If there are three or four different groups bonded to the carbon atoms in the C=C group, the *E/Z* system must be used.

- Work out the priorities of the four groups (the higher the atomic numbers of the elements in the group, the higher the priority.
- If the higher priority groups are on opposite sides, it is the *E*-isomer
- If the higher priority groups are on the same side, it is the *Z*-isomer

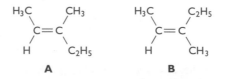

The higher priority on the left-hand carbon is the CH_3 group and the higher on the right carbon is the C_2H_5 group.

Thus compound A is (*E*)-3-methylpent-2-ene, as the two higher priority groups are on opposite sides, and compound B is (*Z*)-3-methylpent-2-ene, as the two higher priority groups are on the same side.

A2 chemistry

Optical isomerism

Optical isomerism is another form of stereoisomerism. It is the result of four different groups being attached to a carbon atom. This carbon atom is called the **chiral centre** and results in **chirality**. A chiral molecule is defined as a molecule that is non-super-imposable on its mirror image.

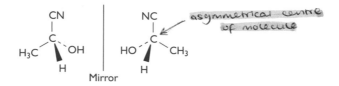

Mirror

Tip You must draw the two isomers as mirror images of each other, with wedges and with dots or dashes to give a three-dimensional appearance to your drawing. Make sure that you do not draw a bond to the wrong atom, for instance to the H of a –CH$_3$, –OH or –COOH group.

Optical isomers can be distinguished from each other because they rotate the plane of plane-polarised light in opposite directions.

Some chemical reactions result in a **racemic** (50:50) mixture of the two optical isomers. Such a mixture does not have any effect on polarised light.

Stereospecificity of nucleophilic substitution of halogenoalkanes

The mechanism for these reactions can be S$_N$1 or S$_N$2 (see pages 25 and 26). If a single optical isomer reacts with a nucleophile, such as hydroxide ions, the type of mechanism can be identified by the effect the final solution has on plane-polarised light.

The S$_N$1 mechanism proceeds through a planar intermediate. This can be attacked either from the top or the bottom and so a racemic mixture containing equal amounts of the two optical isomers will be produced. This will have no effect on the plane of polarisation of the plane-polarised light.

However, in the S$_N$2 mechanism, the nucleophile attacks from the side opposite to the halogen. This will result in the production of a single optical isomer (with inverted configuration), and so the solution will rotate the plane of polarisation of the polarised light.

Carbonyl compounds

Carbonyl compounds contain the C=O group. There are two types of carbonyl compound.

- **Aldehydes** have a hydrogen atom on the carbonyl carbon and so have a –CHO group. Examples include methanal (HCHO), ethanal (CH_3CHO) and propanal (CH_3CH_2CHO).
- **Ketones** have *two* alkyl groups attached to the carbonyl carbon atom. Examples include propanone (CH_3COCH_3) and butanone ($CH_3COCH_2CH_3$).

Physical properties

Neither aldehydes nor ketones have a hydrogen atom that is sufficiently $\delta+$, so they cannot form intermolecular hydrogen bonds. This means that they have a lower boiling point than alcohols with the same number of electrons.

However, they contain a $\delta-$ oxygen atom, and its lone pair of electrons can form a hydrogen bond with the $\delta+$ hydrogen atoms in water. Thus lower members of both aldehyde and ketone homologous series are water-soluble. The solubility decreases as the number of carbon atoms increases due to the hydrophobic hydrocarbon tail.

Preparation

Aldehydes

An aldehyde is prepared by oxidising a *primary* alcohol and distilling off the aldehyde as it is produced. For example, a mixture of potassium dichromate(VI) and sulfuric acid is added to hot ethanol and the ethanal distils off.

$$CH_3CH_2OH + [O] \longrightarrow CH_3CHO + H_2O$$

Ketones

Ketones are prepared by heating a *secondary* alcohol under reflux with an oxidising agent, such as a solution of potassium dichromate(VI) and sulfuric acid.

$$CH_3CH(OH)CH_3 + [O] \longrightarrow CH_3COCH_3 + H_2O$$

Note: in both these reactions, the colour changes from orange ($Cr_2O_7^{2-}$) to green (Cr^{3+}). Tertiary alcohols are not oxidised, so the solution would stay orange.

Reactions of aldehydes and ketones

The carbon atom is $\delta+$. Therefore, carbonyl compounds react with nucleophiles.

Reaction with 2,4-dinitrophenylhydrazine

Both aldehydes and ketones form an orange-yellow precipitate with a solution of 2,4-dinitrophenylhydrazine.

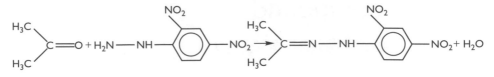

This is an example of a **nucleophilic substitution** reaction. It is used as a test to identify aldehydes and ketones.

Reaction with hydrogen cyanide in the presence of a catalyst of cyanide ions

Although the reactant is hydrogen cyanide, potassium cyanide in a buffer solution at pH 8 is added to the carbonyl compound. This is necessary because the attack is by the nucleophile CN^-, followed by the addition of H^+. For the reaction with ethanal, the organic product is 2-hydroxypropanenitrile.

$$CH_3CHO + HCN \longrightarrow CH_3CH(OH)CN$$

For the reaction with propanone, the organic product is 2-hydroxy-2-methylpropanenitrile.

$$CH_3COCH_3 + HCN \longrightarrow CH_3C(CN)(OH)CH_3$$

The overall reactions are examples of nucleophilic addition.

The mechanism of this reaction, using ethanal as an example, is:

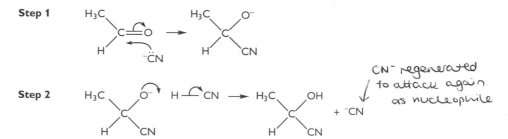

CN⁻ regenerated to attack again as nucleophile

Stereospecificity of nucleophilic addition of HCN to carbonyl compounds

Because the original carbonyl compound is planar at the reaction site, the cyanide ion can attack from either above or below. This means that a racemic mixture of the two enantiomers is formed when aldehydes or asymmetric ketones react.

$$CH_3CHO + HCN \longrightarrow \text{racemic mixture of}$$

and

$$CH_3COC_2H_5 + HCN \longrightarrow \text{racemic mixture of}$$

and

Tip Do not confuse this reason with that involving S_N1 mechanisms. The addition to carbonyl compounds results in a racemic mixture because the *starting* substance is planar, whereas the racemic mixture is formed in S_N1 because the *intermediate* is planar.

Reduction

Aldehydes and ketones can be reduced by lithium aluminium hydride ($LiAlH_4$) in dry ether, *followed* by hydrolysis of the intermediate by dilute acid.

$$CH_3COCH_3 + 2[H] \longrightarrow CH_3CH(OH)CH_3$$

This is another example of a nucleophilic addition reaction.

Reactions limited to aldehydes

Oxidation reactions

Aldehydes contain the –CHO group, which can be oxidised to –COOH or to the –COO⁻ ion. Ketones do not have an easily removed hydrogen atom and so are *not* oxidised by the reagents below.

Reaction with Fehling's or Benedict's solution

When an aldehyde is warmed with Fehling's solution (or Benedict's solution), a red precipitate of copper(I) oxide is formed from the blue solution. The aldehyde is oxidised to a carboxylate ion.

$$CH_3CHO + [O] + OH^- \longrightarrow CH_3COO^- + H_2O$$

Ketones have no effect on Fehling's or Benedict's, which stay blue. This is a test that can be used to distinguish between aldehydes and ketones.

Reaction with Tollens' reagent (alkaline ammoniacal silver nitrate)

When an aldehyde is added to Tollens' reagent and warmed, a silver mirror is formed. Ethanal is oxidised to the ethanoate ion.

$$CH_3CHO + [O] + OH^- \longrightarrow CH_3COO^- + H_2O$$

Ketones do not form a silver mirror, and so this test can also be used to distinguish between an aldehyde and a ketone.

Reaction with acidified dichromate ions

When an aldehyde is warmed with a solution of potassium dichromate(VI) in sulfuric acid, the orange colour turns green as the aldehyde is oxidised to a carboxylic acid.

$$CH_3CHO + [O] \longrightarrow CH_3COOH$$

Iodoform reaction of ethanal and methyl ketones

Compounds with a –$COCH_3$ group undergo this reaction when gently warmed with iodine and a few drops of alkali. A pale yellow precipitate of iodoform, CHI_3, is formed.

$$CH_3CHO + 3I_2 + 4NaOH \longrightarrow CHI_3 + 3NaI + HCOONa + 3H_2O$$

$$CH_3COR + 3I_2 + 4NaOH \longrightarrow CHI_3 + 3NaI + RCOONa + 3H_2O$$

Note: ethanol (CH_3CH_2OH) and secondary alcohols containing the $CH_3CH(OH)$ group also undergo this reaction. They are initially oxidised to ethanal and methyl ketones respectively, which react further to produce iodoform.

Carboxylic acids

Carboxylic acids contain the –COOH group.

Their general formula is $C_nH_{2n+1}COOH$, or RCOOH, where R is an alkyl group.

They are all weak acids, and so they are only slightly ionised in water.

$$RCOOH + H_2O \longrightarrow H_3O^+ + RCOO^-$$

$$K_a = \frac{[H_3O^+][RCOO^-]}{[RCOOH]}$$

Physical properties

Their boiling temperature is higher than both aldehydes and alcohols with the same number of carbon atoms. Hydrogen bonding occurs between the $\delta+$ hydrogen of the OH group and the lone pair of electrons on the oxygen of the C=O group in another molecule. This is stronger than in alcohols as dimers are formed.

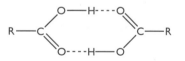

Acids are also able to form hydrogen bonds with water, so the lower members of the homologous series are water-soluble.

Preparation

Oxidation of primary alcohols

When a primary alcohol (or an aldehyde) is heated under reflux with a solution of potassium dichromate(VI) in sulfuric acid, it is oxidised to a carboxylic acid.

$$CH_3CH_2CH_2OH + 2[O] \longrightarrow CH_3CH_2COOH + H_2O$$

Hydrolysis of nitriles

When a nitrile is heated under reflux with a dilute acid such as hydrochloric acid, it is hydrolysed to a carboxylic acid.

$$CH_3CH_2CN + 2H_2O + HCl \longrightarrow CH_3CH_2COOH + NH_4Cl$$

The hydroxynitrile, $CH_3CH(OH)CN$, formed from the reaction of ethanal with hydrogen cyanide can be hydrolysed to 2-hydroxypropanoic acid (lactic acid).

$$CH_3CH(OH)CN + 2H_2O + HCl \longrightarrow CH_3CH(OH)COOH + NH_4Cl$$

Reactions of ethanoic acid, CH_3COOH

Reaction with bases
- Ethanoic acid reacts with aqueous sodium hydroxide to give a salt, sodium ethanoate.
$$CH_3COOH + NaOH \longrightarrow CH_3COO^-Na^+ + H_2O$$
- It reacts with sodium carbonate (solid or in solution), giving off bubbles of carbon dioxide.
$$2CH_3COOH + Na_2CO_3 \longrightarrow 2CH_3COO^-Na^+ + H_2O + CO_2$$
- It reacts with aqueous sodium hydrogen carbonate, giving off bubbles of carbon dioxide.
$$CH_3COOH + NaHCO_3 \longrightarrow CH_3COO^-Na^+ + H_2O + CO_2$$

This reaction is a test for an acid.

Reaction with phosphorus(v) chloride (phosphorus pentachloride)
If phosphorus(v) chloride is added to a dry sample of ethanoic acid, steamy fumes of hydrogen chloride are given off and an acid chloride is produced. The organic product is ethanoyl chloride.

$$CH_3COOH + PCl_5 \longrightarrow CH_3COCl + HCl + POCl_3$$

Note: alcohols (and water) also give steamy fumes with phosphorus(v) chloride, as this is a reaction of the OH group.

Reduction with lithium aluminium hydride
Carboxylic acids are reduced to primary alcohols by lithium aluminium hydride. This is a two-stage process. The first stage is to add lithium aluminium hydride to a solution of the organic acid in dry ether. The second is to hydrolyse the adduct formed with aqueous acid. The reducing agent is represented by [H] in the overall equation:

$$CH_3COOH + 4[H] \longrightarrow CH_3CH_2OH + H_2O$$

Reaction with alcohols — esterification
When ethanoic acid is warmed under reflux with an alcohol, in the presence of a few drops of concentrated sulfuric acid catalyst, an ester is formed. With methanol, methyl ethanoate is produced.

$$CH_3COOH + CH_3OH \rightleftharpoons CH_3COOCH_3 + H_2O$$

If the products of this reaction are poured into cold water, the characteristic sweet smell of an ester can be detected. Many simple esters are used as artificial food flavourings.

Acid chlorides

Acid chlorides contain the group:

The carbon atom of the C=O group is δ+. Therefore, acid chlorides react with nucleophiles.

Reactions

Reaction with water
Acid chlorides are hydrolysed by water. The carboxylic acid is produced — for example:

$$CH_3COCl + H_2O \longrightarrow CH_3COOH + HCl$$

Reaction with alcohols
On mixing, a rapid reaction takes place at room temperature and an ester is formed. For example, ethanoyl chloride reacts with propan-1-ol to form 1-propylethanoate.

$$CH_3COCl + CH_3CH_2CH_2OH \longrightarrow CH_3COOCH_2CH_2CH_3 + HCl$$

This is a rapid and complete reaction, unlike the slow, reversible esterification of a carboxylic acid with an alcohol.

Reaction with ammonia
With concentrated ammonia solution, an amide is rapidly formed. For example, with ethanoyl chloride the product is ethanamide.

$$CH_3COCl + 2NH_3 \longrightarrow CH_3CONH_2 + NH_4Cl$$

Reaction with amines
In this reaction, a substituted amide is produced, which contains the –CONH– peptide link — for example:

$$CH_3COCl + C_2H_5NH_2 \longrightarrow CH_3CONHC_2H_5 + HCl$$

Esters

Esters contain the group:

Reactions

Hydrolysis with acids
Esters are *reversibly* hydrolysed when heated under reflux with an aqueous solution of a strong acid, which acts as a catalyst. Ethyl ethanoate is hydrolysed to ethanoic acid and ethanol.

$$CH_3COOC_2H_5 + H_2O \rightleftharpoons CH_3COOH + C_2H_5OH$$

Reaction with alkalis

When an ester is heated under reflux with an aqueous solution of an alkali such as sodium hydroxide, the ester is hydrolysed to an alcohol and the salt of a carboxylic acid in an *irreversible* reaction.

$$CH_3COOCH_3 \quad + \quad NaOH \quad \longrightarrow \quad CH_3OH \quad + \quad CH_3COONa$$

     methyl ethanoate   methanol   sodium ethanoate

Fats are esters of propane-1,2,3-triol (glycerol) and large carboxylic acids such as stearic acid ($C_{17}H_{35}COOH$). When heated under reflux with aqueous sodium hydroxide, sodium stearate ($C_{17}H_{35}COONa$) is produced. This is a soap and the reaction is called **saponification.**

$$CH_2OOCC_{17}H_{35}$$
$$|$$
$$CHOOCC_{17}H_{35} \quad + 3NaOH \quad \rightarrow \quad CH_2(OH)CH(OH)CH_2OH + 3C_{17}H_{35}COONa$$
$$|$$
$$CH_2OOCC_{17}H_{35}$$

Transesterification reactions

When an ester is reacted with an alcohol in the presence of an acid catalyst, a transesterification reaction takes place.

$$CH_3COOC_2H_5 \quad + \quad CH_3OH \quad \rightleftharpoons \quad CH_3COOCH_3 \quad + \quad C_2H_5OH$$

   ethyl ethanoate   methanol   methyl ethanoate   ethanol

A similar reaction takes place with a carboxylic acid.

$$CH_3COOC_2H_5 \quad + \quad HCOOH \quad \rightleftharpoons \quad HCOOC_2H_5 \quad + \quad CH_3COOH$$

   ethyl ethanoate   methanoic acid   ethyl methanoate   ethanoic acid

Use of this reaction is made in the manufacture of biodiesel and of low-fat spreads.

- Biodiesel — Vegetable oils are esters of propane-1,2,3-triol (glycerol) and mainly unsaturated acids. They are too viscous to be a good fuel, so they are reacted with methanol in a transesterification reaction.

     ester of propane-1,2,3-triol + $3CH_3OH \longrightarrow$

                             three esters of methanol + propane-1,2,3-triol

- Low-fat spreads — Vegetable oils can be hardened to make margarine in two ways. One is partial hydrogenation, but this converts some of the isomers into the *trans* form, which is bad for health. The other alternative is to carry out a partial transesterification reaction with a saturated acid such as stearic acid, $C_{17}H_{35}COOH$. This replaces one of the unsaturated acids in the propane-1,2,3-ester and so increases the melting temperature.

Polyesters

These are formed by the reaction between monomers with two functional groups.

The most common is between a monomer with two COOH groups and one with two OH groups.

Terylene is a polyester made from benzene-1,4-dicarboxylic acid and ethane-1,2-diol.

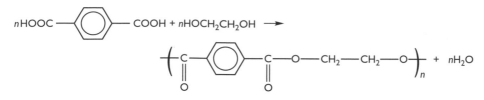

Tip Polyesters have a repeat unit with *four* oxygen atoms — one at each end and two in the middle.

Biopol is a polyester which is biodegradable and is made from 3-hydroxybutanoic acid, $CH_3CH(OH)CH_2COOH$. This monomer has one OH group and one COOH group and so can form a polyester with other 3-hydroxybutanoic acid molecules.

A fragment showing *two* repeat units is:

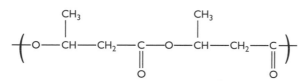

Spectroscopy and chromatography

Spectroscopy

Electromagnetic radiation

The range that needs to be considered for A-level is shown below:

Frequency ⟶

Radio waves $\approx 10^6$ Hz	Microwaves $\approx 10^{11}$ Hz	Infrared $\approx 10^{13}$ Hz	Visible $\approx 10^{14}$ Hz	Ultraviolet $\approx 10^{15}$ Hz

Frequency is measured in hertz, symbol Hz, and is proportional to wave number $\frac{1}{\lambda}$ which has units of cm^{-1}.

Electromagnetic radiation at different frequencies has a number of effects on atoms or molecules.

- Radio waves can affect the way in which nuclei, such as hydrogen nuclei, spin in a strong magnetic field.
- Microwaves cause molecules with polar bonds to rotate faster.
- Infrared radiation causes bending and stretching of bonds, as long as this results in a change of the dipole moment of the molecule.
- Visible light can promote electrons to a higher energy level, causing the substance to be coloured (see Unit 5).
- Ultraviolet radiation can cause covalent bonds to break homolytically, forming radicals.

Radio waves and NMR spectroscopy

Hydrogen nuclei are spinning. This results in the nuclei having a weak magnetic field. When they are placed in a strong magnetic field, the nucleus's own magnetic field can lie parallel with the applied magnetic field or at right angles to it. There is a slight energy difference between these two states. The nuclei can then absorb radio waves and move from the lower (parallel) state to the higher energy (antiparallel) state.

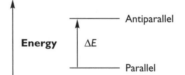

The frequency of radio waves absorbed depends upon the chemical environment of the hydrogen atoms in the molecule. The amount that this differs from a reference substance is called the **chemical shift**, δ.

Low-resolution NMR spectra

These appear as peaks due to the various chemical environments of the hydrogen nuclei in the molecule.

Thus propanone (CH_3COCH_3) has a single peak, as the chemical environments of the hydrogen nuclei in the two CH_3 groups are identical.

Propanal (CH_3CH_2CHO) has three peaks, one due to the CH_3 hydrogens, one due to the CH_2 hydrogens and one due to the CHO hydrogen. These peaks have heights in the ratio of 3:2:1, as there are three H nuclei in the CH_3 group, two in the CH_2 group and just one in the CHO group.

High-resolution NMR spectra

The spectra produced by modern machines are more detailed. The magnetic fields of hydrogen atoms on neighbouring *carbon* atoms slightly alter the applied magnetic field felt by the hydrogen nucleus. This is called spin coupling and causes a splitting of the peak according to the $(n + 1)$ rule.

- If there is *one* hydrogen atom on a neighbouring carbon atom, the peak is split into (1 + 1) = *two* peaks.
- The peak due to the CH_3 hydrogen nuclei is split into two by the CH hydrogen in propan-2-ol, $CH_3CH(OH)CH_3$.
- The peak due to the hydrogen nucleus in the CH group of propan-2ol, $CH_3CH(OH)CH_3$, is split into *seven* as it has six hydrogen atoms on neighbouring carbon atoms (three in each CH_3 group).
- If there are *two* hydrogen atoms on neighbouring carbon atoms, the peak is split into *three* and, if there are three neighbouring hydrogen atoms, the peak is split into *four*.
- The peak due to the CH_3 hydrogen nuclei in ethanol (CH_3CH_2OH) is split into three by the two hydrogen nuclei in the CH_2 group and the peak due to the CH_2 hydrogen nuclei is split into four.
- If there are no neighbouring hydrogen atoms, the peak is not split, thus the peak due to the CH_3 groups in propanone (CH_3COCH_3) is not split.
- The peak due to the hydrogen atom of an OH group is never split and has no effect on other hydrogen nuclei.
- The unsplit peak in the NMR spectrum of ethanol is due to the OH hydrogen nucleus.

Propan-2-ol

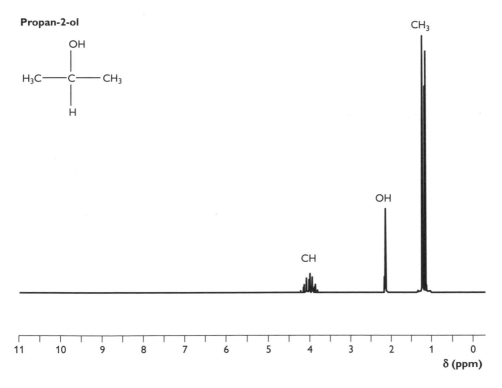

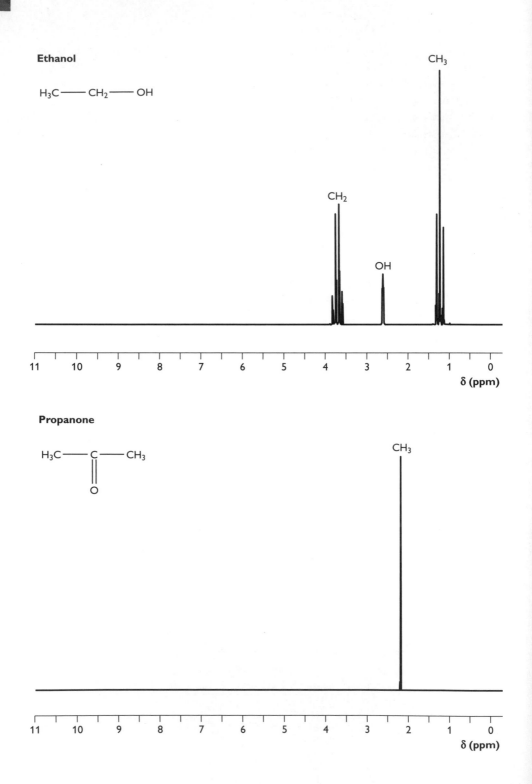

Ethanol

H_3C —— CH_2 —— OH

Propanone

H_3C —— $\overset{\displaystyle C}{\underset{\displaystyle O}{\|}}$ —— CH_3

The same technique is used to study the different environments of water in soft human tissue, such as the brain, spinal discs or cartilage in joints. This is called magnetic resonance imaging or MRI.

Microwaves

Polar molecules such as water and ethanol absorb microwaves. This causes the molecules to rotate faster and when they collide with other molecules, this rotational energy is converted into kinetic energy — heat. Domestically, use is made of this in microwave ovens to heat food and drinks.

In the pharmaceutical industry, microwave ovens are used to heat reaction mixtures. This is very energy efficient and there is no local overheating which could cause decomposition of fragile molecules.

Infrared spectroscopy

An infrared spectrum is obtained by passing infrared radiation with a range of frequencies through a substance. The different bonds in the substance absorb radiation at different frequencies, causing the bond to stretch or bend. This results in a series of peaks or lines, each representing different bonds.

Broad bands are produced by O–H and N–H bonds, because of hydrogen bonding.

Approximate frequencies, measured in wave numbers, are shown in the following table.

Wave number / cm^{-1}	Group
1700	C=O
3300	O–H in alcohols
3000	O–H in acids
3400	N–H
2900	C–H
1420	C–C in alkanes
1655	C=C in alkenes

For a fuller list consult the Edexcel data booklet or other sources.

You will normally have access to a table of frequencies and bonds. However, you ought to remember that:
- a sharp band around 1700 cm^{-1} is caused by C=O in aldehydes, ketones, acids and esters
- a broad band around 3300 cm^{-1} is caused by O–H in alcohols

The IR spectra below are of propan-1-ol and its oxidation product propanal. The propan-1-ol spectrum shows the broad alcohol O–H peak at 3300 cm^{-1} and no peak around 1700 cm^{-1}. The propanal spectrum shows a sharp C=O peak at 1720 cm^{-1} and no peak around 3300 cm^{-1}.

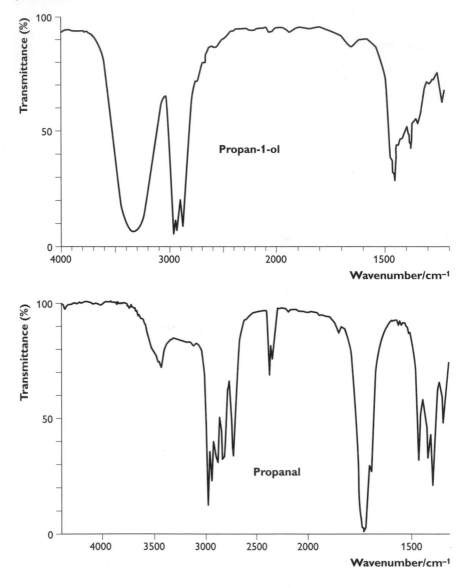

Following a reaction

Infrared spectroscopy can be used to follow a reaction. For example the reduction of 3-hydroxypropene (CH$_2$=CH-CH$_2$OH) to propan-1-ol can be studied by watching the band at around 1655 cm^{-1}, due to the C=C group, disappear.

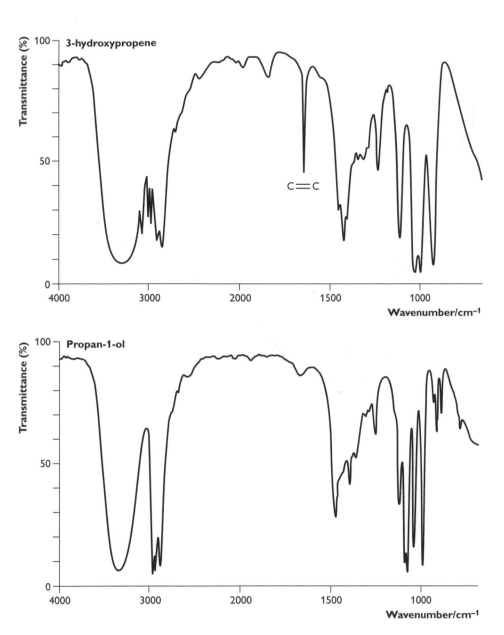

Ultraviolet radiation

The high energy of this type of radiation can cause covalent bonds to break. The reaction between an alkane and chlorine is triggered by UV light. This causes the homolytic fission of the Cl–Cl bond. The chlorine radicals produced then cause a chain reaction.

Initiation
$$Cl_2 \xrightarrow{UV} 2Cl\bullet$$

Propagation
$$Cl\bullet + CH_4 \longrightarrow HCl + CH_3\bullet$$

$$CH_3\bullet + Cl_2 \longrightarrow CH_3Cl + Cl\bullet$$

Chain breaking
$$Cl\bullet + Cl\bullet \longrightarrow Cl_2$$

$$Cl\bullet + CH_3\bullet \longrightarrow CH_3Cl$$

$$CH_3\bullet + CH_3\bullet \longrightarrow C_2H_6$$

Mass spectra

Mass spectrometer

In a mass spectrometer, molecules are bombarded with high-energy electrons. Positively charged ions are formed, which give rise to peaks in the mass spectrum. These peaks result from:

- the molecular ion, M^+, of mass/charge ratio (m/e) equal to the relative molecular mass of the substance. This is the ion with the *largest m/e* value in the spectrum.
- ionic fragments caused by the break-up of the molecular ion. The radicals lost provide information about the structure of the molecule under test.

m/e units lost	Group lost
15	CH_3
29	C_2H_5 or CHO
31	CH_2OH

The mass spectrum of propan-1-ol is shown below.

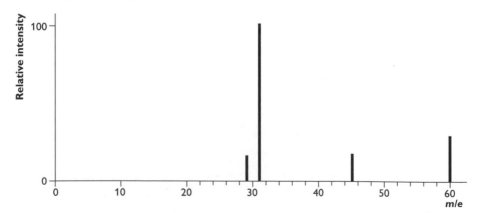

The spectrum details are summarised in the following table.

Peak at *m/e*	Ionic fragment	Group lost
60	$(CH_3CH_2CH_2OH)^+$	None
45	$(CH_2CH_2OH)^+$	CH_3
31	$(CH_2OH)^+$	C_2H_5
29	$CH_3CH_2^+$	CH_2OH

Propan-2-ol would give a peak at $m/e = 45$ due to loss of CH_3 but not at $m/e = 29$ or $m/e = 31$ because it does not have a C_2H_5 group or a CH_2OH group.

Chromatography

All forms of chromatography have a stationary phase. This can be a solid packed into a column, a thin layer of solid on a support or a liquid adsorbed on a solid matrix.

The mixture is then passed through the stationary phase with an eluent, which is either a liquid or a gas.

The different components travel through the stationary phase at different speeds and so are separated.

Gas–liquid chromatography

Gas–liquid chromatography involves a sample of a liquid mixture being injected into the top of a chromatographic column that is in a thermostatically controlled oven. The sample evaporates and is forced through the column by the flow of nitrogen gas or another inert, gaseous eluent. The column itself contains a liquid stationary phase that is adsorbed onto the surface of an inert solid. The gases coming out pass through a detector — usually a device that measures the thermal conductivity of the gas.

The rate at which a substance passes through the column depends on the extent to which it interacts with the liquid stationary phase.

The identity of the components in the mixture can be found by removing samples leaving the column and measuring their infrared spectra. These are then compared with spectra in a database.

High-pressure liquid chromatography (HPLC)

This consists of a column packed with a solid of uniform particle size — the stationary phase. The sample to be separated is dissolved in a suitable solvent and added at the top of the column. The liquid eluent — the moving phase — is then forced through the column under high pressure.

Different substances have different strengths of interaction between the stationary phase and the moving phase. The time taken for a component in the sample mixture to pass through the column is called the retention time and is a unique characteristic of the substance, the composition of the eluent, the nature of the stationary phase and the pressure. This means that different components will pass through one after the other with gaps between each.

The use of high pressure increases the speed at which the eluent passes through the column and so reduces the extent to which the band of a component spreads out due to diffusion. This gives it a much higher resolution than paper or thin layer chromatography. The column can be connected to an infrared spectrometer, which can then identify each component in the mixture.

Thin layer chromatography

This is used to separate amino acids and the technique is described in the Unit 5 Guide in the section on amino acids.

Questions
&
Answers

This section contains multiple-choice and structured questions similar to those you can expect to find in Unit Test 4. The questions given here are not balanced in terms of types of questions or level of demand.

Examiner comments

The sample answers are followed by examiner comments, preceded by the icon e. These comments may explain the correct answer or point out common errors.

Multiple-choice questions

Each question or incomplete statement is followed by four suggested answers, A, B, C and D. Select the *best* answer in each case. The answers are given, with some commentary, after Question 15.

1 **Consider the reaction:**

$A + 2B \longrightarrow C + D$

The reaction was studied and the following data on initial rates were obtained.

Experiment	[A]/mol dm^{-3}	[B]/mol dm^{-3}	Rate/mol dm^{-3} s^{-1}
1	0.10	0.10	4.6×10^{-4}
2	0.30	0.10	1.4×10^{-3}
3	0.15	0.20	1.4×10^{-3}

What is the correct rate equation?

A rate $= k[A]$

B rate $= k[A]^3$

C rate $= k[A][B]$

D rate $= k[A][B]^2$

2 **For a specific reaction, the rate equation is:**

rate $= k[X]^2[Y]$

The units of rate are mol dm^{-3} s^{-1}. The units of the rate constant, k, for this reaction are:

A mol^{-1} dm^3 s^{-1}

B mol dm^{-3} s

C mol^{-2} dm^6 s^{-1}

D mol^2 dm^{-6} s

3 **An exothermic reaction is thermodynamically spontaneous at 20°C. Which of the following statements is correct?**

A It will be spontaneous at all temperatures whatever the sign of ΔS_{system}.

B It will only proceed at 20°C if the activation energy is not too high.

C It will only proceed if ΔS_{system} is positive.

D The value of ΔS_{total} is larger at 40°C than at 20°C.

4 **Which of the following does NOT have a positive value of ΔS_{system}?**

A $MgSO_4(s) + aq \longrightarrow Mg^{2+}(aq) + SO_4^{2-}(aq)$

B $Ba(OH)_2.8H_2O(s) + 2NH_4Cl(s) \longrightarrow BaCl_2(s) + 10H_2O(l) + 2NH_3(g)$

C $NaCl(s) \longrightarrow NaCl(l)$

D $C(s) + H_2O(g) \longrightarrow CO(g) + H_2(g)$

5 Consider the equilibrium reaction:

$$A(g) + B(g) \rightleftharpoons C(g) \qquad \Delta H = +23 \text{ kJ mol}^{-1}$$

Which of the following statements is true?

A An increase in pressure will drive the position of equilibrium to the right and hence will increase the value of the equilibrium constant, K.

B An increase in pressure will increase the value of the equilibrium constant, K, and hence drive the position of equilibrium to the right.

C An increase in temperature will drive the position of equilibrium to the right and hence will increase the value of the equilibrium constant, K.

D An increase in temperature will increase the value of the equilibrium constant, K, and hence drive the position of equilibrium to the right.

6 The Haber process for the manufacture of ammonia utilises the exothermic reaction between nitrogen and hydrogen. It is carried out at 400°C at a pressure of 200 atm in the presence of a heterogeneous catalyst of iron. Which of the following statements is **NOT** true about this process?

A High pressure is used to increase the equilibrium yield. $N_2 + 3H_2 \rightleftharpoons 2NH_3$

B High pressure is used to speed up the rate of reaction.

C The value of the equilibrium constant is small at 400°C and is even smaller at 500°C.

D A temperature of 400°C and an iron catalyst are chosen because these conditions result in a reasonable yield at a fast rate.

7 A buffer solution of a weak acid and its salt has a pH of 5.00. When it is diluted ten-fold, its pH:

A stays the same

B alters by a small amount only, because a buffer resists changes in pH.

C falls to 4.00

D increases to 6.00

8 The correct name for the following compound is:

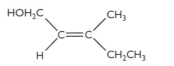

A E-1-hydroxy-3-methylpent-2-ene

B Z-1-hydroxy-3-methylpent-2-ene

C cis-1-hydroxy-3-methylpent-2-ene

D trans-1-hydroxy-3-methylpent-2-ene

9 The compound $CH_3CH(OH)CH(CH_3)COOH$ has:

A no optical isomers

B two optical isomers

C three optical isomers

D four optical isomers

10 In the nucleophilic addition of **HCN** to propanone, the first step is attack by a lone pair of electrons:

A on the oxygen atom of propanone onto the δ+ hydrogen atom of **HCN**

B on the carbon atom of **HCN** onto the δ+ carbon atom in propanone

C on the carbon atom of the **CN**$^-$ ion onto the δ+ carbon atom in propanone

D on the nitrogen atom of the **CN**$^-$ ion onto the δ+ carbon atom in propanone

11 In which of the following is there hydrogen bonding?

A between different molecules of propanone

B between different molecules of propanal

C between propanal and propanone molecules

D between propanal molecules and water

12 Consider the reaction scheme:

$$\text{CH}_3\text{COOH} \xrightarrow{\text{step 1}} \text{CH}_3\text{COCl} \xrightarrow{\text{step 2}} \text{CH}_3\text{CONH}_2$$

What are the correct reagents for steps 1 and 2?

	Step 1	Step 2
A	PCl_5	KCN
B	PCl_5	NH_3
C	HCl	KCN
D	HCl	NH_3

13 The correct repeat unit for the polyester formed by reacting hexane-1,6-dioic acid with hexane-1,6-diol is:

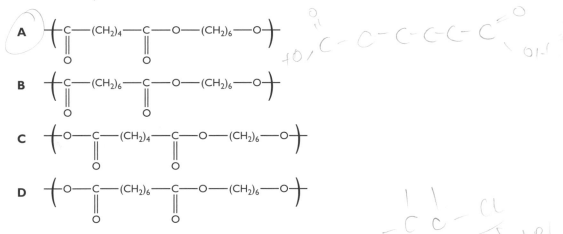

14 The NMR spectra of CH_3CH_2Cl and CH_3CH_2Br:

A both have two peaks in the ratio 3:2 at the same chemical shift

B both have two peaks in the ratio 3:2 at different chemical shifts

C both have two peaks split into 2 and 3

D both have three peaks in the ratio 3:2:1 due to the C–H bond in CH_3, the C–H bond in CH_2 and the C–halogen bond

15 The mass spectrum of CH_3CH_2Br will have:

A a peak at $m/e = 109$

B peaks at $m/e = 108$ and 110

C peaks at $m/e = 93, 94$ and 95

D peaks at $m/e = 79, 80$ and 81

■ ■ ■

Answers

1 C

When experiments 1 and 2 are compared, [A] increases by a factor of 3 and the rate increases by a factor of $1.4 \times 10^{-3}/4.6 \times 10^{-4} = 3.0$, so it is first order with respect to A. Consider experiments 2 and 3: [A] halves and [B] doubles, but the rate stays the same. The halving of [A] would reduce the rate by a factor of 2, but as the rate did not alter, the doubling of [B] must have increased the rate by a factor of 2, cancelling out the decrease due to the lower concentration of A. This means that the reaction is first order with respect to B.

2 C

The rate equation has to be rearranged into:

$$k = \frac{\text{rate}}{[X]^2[Y]}$$

which, on substituting, becomes:

$$k = \frac{\text{concentration} \times \text{time}^{-1}}{\text{concentration}^2 \times \text{concentration}} = \text{concentration}^{-2}\,\text{time}^{-1} = \text{mol}^{-2}\,\text{dm}^6\,\text{s}^{-1}$$

In options **B** and **D** the expression for k is upside down; options **A** and **B** do not show that it is second order in [X].

3 B

For a thermodynamically feasible reaction to be observed, the activation energy must be reasonably low, so that the reactants are kinetically unstable as well as being thermodynamically unstable. Thus **B** is the correct response. It is worth checking the others. For a reaction to be spontaneous, ΔS_{total} must be positive. Since the reaction is exothermic, ΔS_{surr} is positive. If ΔS_{system} is negative, the reaction will cease to be spontaneous as the temperature is increased, because ΔS_{surr} will become less positive. This means that **A** is false. ΔS_{system} need not be positive. If it is negative and numerically less than the positive ΔS_{surr}, the reaction is feasible. Thus option **C** is false. ΔS_{total} is not larger at $40°C$. It is an exothermic reaction and so the value of ΔS_{total} (and hence the equilibrium constant K) will decrease with an increase in temperature. Thus option **D** is also false.

4 A

Even though there is a large increase in entropy of the solute going from solid to aqueous solution, the water is ordered by the high charge density of the group 2 cation. This more than outweighs the increase in entropy of the magnesium sulfate. Therefore ΔS_{system} is negative, making option **A** the correct answer to this negative question. It is worth putting '+' or '−' beside each response. Then the one marked '−' will be your answer. In **B**, a highly disordered gas is produced, so ΔS_{system} is positive. A molten liquid always has a higher entropy than its solid, so option **C** is positive. In **D**, 2 moles of gas are formed from 1 mole of gas, which causes an increase in entropy. Therefore, **D** is positive. The only answer with a negative ΔS_{system} is **A**.

5 **D**

A change in pressure will not alter the value of the equilibrium constant, K, so options **A** and **B** cannot be correct. The logic in option **C** is wrong, but is correct in option **D**. For any endothermic reaction, an increase in temperature will cause an increase in the value of K, which becomes bigger than the equilibrium term $[products]^n/[reactants]^m$. This term has to get larger, so that it once again equals the value of K. To do this, the value of the upper line must increase and that of the bottom line decrease. This happens as the position of equilibrium shifts to the right, so the answer is **D**. The logic for an *endo*thermic reaction is:

T increases, therefore K increases, therefore the equilibrium position moves to the right and is **not**:

T increases, therefore the equilibrium position moves to the right, so K increases.

6 **B**

This question is best answered by putting *true* or *false* beside each option:
- Option A: True — there are fewer gas moles on the right, so an increase in pressure will drive the equilibrium to the right and so increase the yield.
- Option B: False — the rate of a reaction involving a solid metallic catalyst at a given temperature is not altered by pressure. It depends solely on the number of active sites on the surface of the catalyst.
- Option C: True — the value of K is small at 400°C (which is why it is economic to use high pressure). Since the reaction is exothermic, the value of K will decrease as the temperature rises.
- Option D: True — a catalyst allows the reaction to proceed at a fast rate at a lower temperature. This results in a higher yield than if a higher temperature were used.

Options **A**, **C** and **D** are all true. Option **B** is false and is the correct answer to this negative question.

7 **A**

$[H^+] = \dfrac{K_a\,[acid]}{[salt]}$, so the pH of a buffer depends on the ratio of [acid] to [salt].

This ratio is not altered by dilution, so the pH stays the same. Option **B** shows a misunderstanding of the term 'buffer'. A buffer resists change in pH when a small amount of *acid* or *alkali* is added. Options **C** and **D** would be true if the ratio of [acid] to [salt] were to increase or decrease by a factor of 10.

8 A

As there are four different groups around the C=C, the *cis/trans* method of naming geometric isomers cannot be used, so options **C** and **D** are incorrect. The group on the right-hand carbon of the C=C that has the higher priority is the C_2H_5 group; that with the higher priority on the left-hand carbon is the CH_2OH group. These are on opposite sides of the double bond, so the compound is an *E*-isomer.

9 D

The compound has two different chiral centres, as shown by the asterisks * on the carbon atoms in the formula:

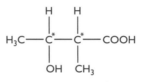

The four isomers are $++, --, +-$ and $-+$. The last two are different because the four groups around the first chiral carbon atom are different from the four groups around the second chiral carbon atom.

10 C

This reaction requires a catalyst of cyanide ions. The formula of a cyanide ion is usually written as CN^-. However, the charge is on the carbon atom, which has a lone pair of electrons. In the first step, the lone pair attacks the δ^+ carbon atom in propanone, forming a new carbon–carbon bond.

11 D

For hydrogen bonding to occur there must be a δ^+ hydrogen atom in one molecule and a δ^- oxygen atom in the other molecule. Propanal has a δ^- oxygen (caused by the considerable difference in electronegativities of the carbon and the oxygen in the C=O group) and there are δ^+ hydrogen atoms in the water. This means that hydrogen bonding will occur. Neither propanone nor propanal has a hydrogen atom that is sufficiently δ^+ for hydrogen bonding, so there are no hydrogen bonds between propanone molecules (option **A**), between propanal molecules (option **B**) or between propanal and propanone molecules (option **C**).

12 B

A carboxylic acid can be converted to an acid chloride by reaction with phosphorus pentachloride, PCl_5, not by reaction with hydrogen chloride. Thus options **C** and **D** are incorrect. The conversion of an acid chloride to an amide is carried out by reaction with ammonia. Thus, option **A** is incorrect.

13 A

📝 Hexane-1,6-dioic acid has six carbon atoms (including the carboxylic acid carbon atoms), so options **B** and **D** are incorrect as they both have eight carbon atoms in the dioic acid parts of their chains. The polyester forms by loss of an OH from one monomer and an H from the other, so the repeat unit has only four oxygen atoms. This means that options **C** and **D** are incorrect because they have five oxygen atoms in their repeat units.

14 B

📝 Peaks in an NMR spectrum are caused by protons in different chemical environments. The protons of the CH_3 group are in a different environment from those in the CH_2 group, so there are two peaks in the spectra of both molecules. This discounts option **D**. The peaks will be split by spin–spin coupling according to the $(n + 1)$ rule, where n is the number of neighbouring hydrogen atoms. This means that the peak due to the CH_3 group will be split into three and that of the CH_2 into four, so option **C** is incorrect. The different halogen atoms will cause a different δ value (chemical shift), so option **A** is incorrect and **B** is correct.

15 B

📝 Bromine has a relative atomic mass of 80. However, there are no atoms of mass 80 units, as bromine consists of two isotopes of masses 79 and 81 in equal proportions. Thus it will **not** have a molecular ion of mass $2 \times 12 + 5 + 80 = 109$ (option **A**). The molecular masses of $CH_3CH_2{}^{79}Br$ and $CH_3CH_2{}^{81}Br$ are 108 and 110, so there will be peaks at *m/e* 108 and 110 (the correct response **B**). There will also be peaks due to loss of CH_3 at $108 - 15 = 93$ and $110 - 15 = 95$, but no peak at $109 - 15 = 94$, so option **C** is incorrect. There will be peaks at *m/e* $= 79$ and 81 due to $^{79}Br^+$ and $^{81}Br^+$, but no peak at *m/e* $= 80$ because there is no ^{80}Br isotope. Therefore, option **D** is incorrect.

Structured questions

Responses to questions with longer answers may have an examiner's comment preceded by the icon 🄴. Some of the comments highlight common errors made by candidates who produce work of C-grade standard or lower.

Question parts marked with an asterisk* test 'Quality of Written Communication' (QWC).

It is worth reading through the whole question before attempting to answer it.

Question 16

(a) **What is meant by:**
    (i) order of reaction (1 mark)
    (ii) rate of reaction (1 mark)

(b) **1-bromopropane and sodium hydroxide react to form propan-1-ol and sodium bromide. The rate of this reaction was measured using different concentrations of hydroxide ions and 1-bromopropane. The results are shown in the table below.**

Experiment	$[OH^-]$/mol dm^{-3}	[1-bromopropane]/ mol dm^{-3}	Initial rate of reaction/mol dm^{-3} s^{-1}
1	0.25	0.25	1.4×10^{-4}
2	0.125	0.25	7.0×10^{-5}
3	0.50	0.125	1.4×10^{-4}

  *(i) **Deduce the order of reaction with respect to hydroxide ions and to 1-bromopropane. Justify your answer.** (4 marks)
  (ii) **Write the rate equation for this reaction.** (1 mark)
  (iii) **Calculate the value of the rate constant.** (2 marks)
  (iv) **Write a mechanism for this reaction that is consistent with your answer to (ii).** (3 marks)

*(c) **In terms of collision theory, explain how the rate of this exothermic reaction would alter with an increase in temperature.** (4 marks)

Total: 16 marks

■ ■ ■

Answers to Question 16

(a) (i) The sum of the powers to which the *concentrations* of the reactants are raised in the experimentally determined rate equation ✓.

🄴 Do not say that it is the sum of the individual orders, unless you also define individual order.

(ii) The rate of change of concentration of a reactant or of a product with respect to time ✓.

💡 The answer 'It is the amount by which the concentration has changed divided by the time for that change' is also acceptable.

(b) (i) Consider experiments 1 and 2. When [OH⁻] is halved and [1-bromopropane] is kept constant, the rate also halves. Therefore, the order with respect to hydroxide ions is 1 ✓.

Consider experiments 1 and 3. When [OH⁻] is doubled and [1-bromopropane] is halved, the rate does not alter. Since the rate should double because of the doubling of [OH⁻] ✓, the halving of [1-bromopropane] must have caused the rate to halve ✓, cancelling out the increase due to the rise in [OH⁻]. Therefore, the reaction is first order ✓ with respect to 1-bromopropane.

💡 You must make it clear which two experiments you are comparing.

(ii) rate = k[OH⁻][1-bromopropane] ✓

(iii) $k = \dfrac{\text{rate}}{[\text{OH}^-][\text{1-bromopropane}]} = \dfrac{1.4 \times 10^{-4}}{0.25 \times 0.25 \checkmark} = 0.0022\,\text{mol}^{-1}\,\text{dm}^3\,\text{s}^{-1}$ ✓

💡 The units of k can be calculated as $\dfrac{\text{concentration} \times (\text{time})^{-1}}{(\text{concentration})^2}$
= (concentration)⁻¹ × (time)⁻¹

(iv) The rate equation is first order with respect to both reactants, so the mechanism is S_N2. It is attack by the nucleophilic OH⁻ ion involving a transition state:

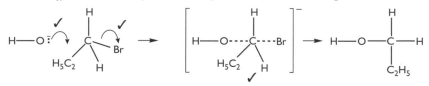

💡 The curly arrow should start from the lone pair of electrons on the oxygen atom of the OH⁻ ion and go towards the δ^+ carbon atom. A second curly arrow must be drawn from the σ C–Br bond to the δ^- bromine atom. The transition state must show a negative charge.

(c) An increase in temperature causes the molecules to gain kinetic energy ✓. As a result, more of the colliding molecules have energy ≥ the activation energy ✓, so a greater *proportion* of the collisions result in reaction ✓ (i.e. more of the collisions are successful). Hence the rate of reaction increases ✓.

💡 For the third mark, do not say that 'There will be more successful collisions'. The total number of successful collisions is independent of the temperature because the reaction will still go to completion, but over a longer time, at a lower temperature.

■ ■ ■

Question 17

Hydrogen can be made by reacting steam with coke at high temperature:

$$C(s) + H_2O(g) \rightleftharpoons CO(g) + H_2(g) \qquad \Delta H = +131 \text{ kJ mol}^{-1}$$

(a) The reaction was allowed to reach equilibrium at a temperature T and at a pressure of 1.2 atm. It was found that 10% of the steam was converted into carbon monoxide and hydrogen.

(i) Write the expression for K_p. (1 mark)

(ii) Calculate the value of K_p at this temperature (to two significant figures). Include the units in your answer. (6 marks)

*(b) State and explain the effect on the equilibrium constant, K_p, and *hence* on the equilibrium yield of hydrogen of:

(i) increasing the pressure to 2.4 atm (4 marks)

(ii) increasing the temperature (3 marks)

(c) Suggest why a temperature of 2000°C is not used. (1 mark)

Total: 15 marks

■ ■ ■

Answers to Question 17

(a) (i) $K_p = \dfrac{p(CO)p(H_2)}{p(H_2O)}$ ✓

🖉 A value for carbon (coke) is not included because it is a solid.

(ii)

	H₂O(g)	CO(g)	H₂(g)	
Initial moles	1	0	0	
Equilibrium moles	$1 - 0.10 = 0.90$	0.10	0.10	✓
Mole fraction	$0.90/1.10 = 0.818$	$0.10/1.10 = 0.0909$	$0.10/1.1 = 0.0909$	✓
Partial pressure/atm	$0.818 \times 1.2 =$ 0.982	$0.0909 \times 1.2 =$ 0.109	$0.0909 \times 1.2 =$ 0.109	✓

$$K_p = \frac{p(CO)p(H_2)}{p(H_2O)} = \frac{0.109\text{ atm} \times 0.109\text{ atm}}{0.982\text{ atm}} = 0.012 \checkmark \text{ atm} \checkmark (2 \text{ s.f.}) \checkmark$$

🖉 The route is:

% ⟶ equilibrium moles ⟶ mole fraction ⟶ partial pressure ⟶ value for K_p ⟶ units of K_p.

(b) (i) Increasing the pressure has no effect on the value of K_p ✓; only temperature change affects this. However, the partial pressure term increases by a factor of two and so is no longer equal to the unaltered K_p ✓. The system is now not at equilibrium and reacts by moving to the left until the partial pressure term once again equals K_p ✓. Thus the equilibrium yield of hydrogen is decreased ✓.

(ii) Increasing the temperature will cause the value of K_p to increase, as the reaction is endothermic ✓. The partial pressure term is now smaller than K_p, so the system is not in equilibrium ✓ and reacts to make more hydrogen and carbon monoxide until the partial pressure term equals the new larger value of K_p ✓.

(c) This temperature would result in greater conversion to hydrogen, but would be uneconomic as water is cheap and such a high temperature is expensive to maintain ✓.

■ ■ ■

Question 18

(a) Propanoic acid, CH_3CH_2COOH, is a weak acid that reacts reversibly with water:

$$CH_3CH_2COOH(aq) + H_2O(l) \rightleftharpoons H_3O^+(aq) + CH_3CH_2COO^-(aq)$$

When an aqueous solution of propanoic acid is mixed with solid sodium propanoate, a buffer solution is formed.

(i) Write the expression for the acid dissociation constant, K_a, of propanoic acid (1 mark)

(ii) Define a buffer solution. (2 marks)

(iii) Calculate the mass of sodium propanoate that has to be added to $100\,cm^3$ of a $0.600\,mol\,dm^{-3}$ solution of propanoic acid to make a buffer solution of pH 5.06 at 25°C. (The acid dissociation constant of propanoic acid = $1.31 \times 10^{-5}\,mol\,dm^{-3}$ at 25°C; molar mass of sodium propanoate = $96.0\,g\,mol^{-1}$.) (4 marks)

*(iv) Explain why a solution of propanoic acid alone is not a buffer solution. (3 marks)

(b) Under suitable conditions, both propanoic acid and propanoyl chloride react with methanol to form an ester.

(i) Name the ester produced. (1 mark)

(ii) Write equations for the formation of this ester from:
■ propanoic acid
■ propanoyl chloride (2 marks)

(iii) Which method gives a better yield of ester. Explain your answer. (1 mark)

(c) Vegetable oils can be converted into biodiesel by a transesterification reaction.
Give one example of this. (2 marks)

(d) Ethane-1,2-diol and benzene-1,4-dicarboxylic acid form a polyester.
Write the structural formula of the repeat unit of this polymer
showing all double bonds. (2 marks)

Total: 18 marks

Answers to Question 18

(a) (i) $K_a = \dfrac{[H_3O^+][CH_3CH_2COO^-]}{[CH_3CH_2COOH]}$ ✓

(ii) A buffer solution is one that maintains a nearly constant pH ✓ when small
amounts of acid or alkali are added ✓.

(iii) $[H_3O^+] = 10^{-pH} = 10^{-5.06} = 8.710 \times 10^{-6}\,mol\,dm^{-3}$ ✓

$[CH_3CH_2COO^-] = \dfrac{K_a[CH_3CH_2COOH]}{[H_3O^+]} = \dfrac{1.31 \times 10^{-5} \times 0.600}{8.710 \times 10^{-6}} = 0.9024\,mol\,dm^{-3}$ ✓

moles of salt = concentration × volume = $0.9024\,mol\,dm^{-3} \times 0.100\,dm^3 =$
0.09024 mol
mass of salt needed = moles × molar mass = 0.09024 × 96.0 = 8.66 g ✓

(iv) The crucial point about a buffer is that the ratio [acid]:[salt] must not alter
significantly when small amounts of acid or alkali are added. This can only
happen if the acid and salt are present in similar amounts. Propanoic acid
is a weak acid and so a solution has a significant $[CH_3CH_2COOH]$ but a tiny
$[CH_3CH_2COO^-]$ ✓. When OH^- ions are added they react with propanoic acid
molecules. This causes $[CH_3CH_2COO^-]$ to rise significantly, which alters the
[acid]:[salt] ratio and affects the pH ✓. When H^+ ions are added, they react with
some of the few $CH_3CH_2COO^-$ ions present, making $[CH_3CH_2COO^-]$ even
smaller. This alters the [acid]:[salt] ratio and causes the pH to change ✓.

(b) (i) Methyl propanoate ✓

(ii) With propanoic acid:
$CH_3CH_2COOH + CH_3OH \rightleftharpoons CH_3CH_2COOCH_3 + H_2O$ ✓
With propanoyl chloride:
$CH_3CH_2COCl + CH_3OH \longrightarrow CH_3CH_2COOCH_3 + HCl$ ✓

(iii) The reaction with propanoyl chloride because it is not a reversible reaction
whereas the reaction with propanoic acid is reversible ✓.

(c) The transesterication reaction involved is when the ester of propane-1,2,3-triol
and unsaturated long-chain fatty acids reacts with an alcohol such as methanol.
This produces three methyl esters of the unsaturated acids, which are used as
biodiesel ✓, and a by-product of propane-1,2,3-triol.

```
CH₃OOCR                                                    CH₂OH
 |                                                          |
CH₃OOCR'  + 3CH₃OH  ⟶  RCOOCH₃ + R'COOCH₃ + R"COOCH₃  +  CHOH ✓
 |                                                          |
CH₃OOCR"                                                   CH₂OH
```

(d)

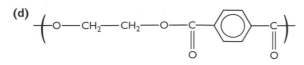

> Ester linkage ✓; rest of repeat unit with 'continuation' bonds ✓. The repeat unit of a polyester will have *four* oxygen atoms.

■ ■ ■

Question 19

Propanone and propanal are isomers with molecular formula C_3H_6O.

(a) Write the equation for the reaction of propanone with 2,4-dinitrophenylhydrazine. (2 marks)

(b) Write the formula of the organic product of the reaction of propanal with:
    (i) Fehling's solution (1 mark)
    (ii) lithium aluminium hydride in dry ether followed by the addition of dilute hydrochloric acid (1 mark)

(c) (i) Both propanone and propanal react with hydrogen cyanide, HCN, in the presence of a catalyst of cyanide ions. Draw the mechanism of the reaction with propanal. (3 marks)
    *(ii) Explain why this reaction is very slow at both high and low pH. (2 marks)

                                                               Total: 9 marks

■ ■ ■

Answers to Question 19

(a)

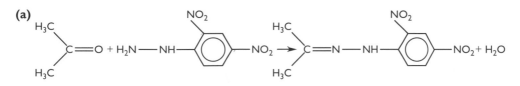

> C=N linkage ✓; rest of molecule ✓.

(b) (i) $CH_3CH_2COO^-$ ✓

> Fehling's solution is alkaline and oxidises propanal to give the anion of propanoic acid, but not the acid itself.

    (ii) $CH_3CH_2CH_2OH$ ✓

(c) (i) Step 1

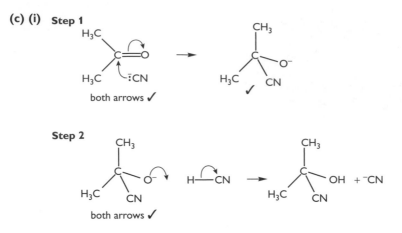

both arrows ✓

Step 2

both arrows ✓

(ii) At a high pH (very alkaline), the OH^- ions react with the HCN. As the [HCN] is now very small, the rate of step 2 is almost zero ✓. At a low pH (very acid), the H^+ ions protonate any CN^- ions. As the $[CN^-]$ is now very small, the rate of step 1 is almost zero ✓.

Data question

Question 20

(a) Consider the reaction:

$$CaSO_4(s) \longrightarrow CaO(s) + SO_3(g)$$

Use the data booklet to calculate:

(i) ΔS_{system} (2 marks)

(ii) ΔH for the reaction and hence ΔS_{surr} at 500°C (3 marks)

(iii) Use your answers to (i) and (ii) to calculate ΔS_{total} at 500°C and hence the value of the equilibrium constant, K, at this temperature.
($R = 8.31\,J\,K^{-1}\,mol^{-1}$) (3 marks)

(iv) Calculate the value of the equilibrium constant at 1000°C (1 mark)

(b) ■ The lattice energy of calcium sulfate, $CaSO_4$, is $-2677\,kJ\,mol^{-1}$.
■ The hydration enthalpy of the $Ca^{2+}(g)$ ion is $-1650\,kJ\,mol^{-1}$; the hydration enthalpy of the $SO_4(g)$ ion is $-1045\,kJ\,mol^{-1}$.

Draw a Hess's law diagram for the dissolving of calcium sulfate in water and calculate its enthalpy of solution. (4 marks)

*(c) Which of strontium sulfate and barium sulfate is more soluble? Justify your answer using the data below to calculate the difference in ΔS_{total}.
■ $\Delta H_{solution}\ SrSO_4(s) = -9\,kJ\,mol^{-1}$
■ $\Delta H_{solution}\ BaSO_4(s) = +19\,kJ\,mol^{-1}$
■ $S\ Sr^{2+}(aq) = -33\,J\,K^{-1}\,mol^{-1}$
■ $S\ Ba^{2+}(aq) = +10\,J\,K^{-1}\,mol^{-1}$ (5 marks)

Total: 18 marks

■ ■ ■

Answers to Question 20

(a) (i) $\Delta S_{system} = S(CaO) + S(SO_3) - S(CaSO_4)$✓
$= +40 + 256 - (+107) = +189\,J\,K^{-1}\,mol^{-1}$ ✓

(ii) $\Delta H = \Delta H_f\,CaO + \Delta H_f\,SO_3 - \Delta H_f\,CaSO_4 = (-635) + (-396) - (-1434) = +403\,kJ\,mol^{-1}$ ✓

$\Delta S_{surr} = -\Delta H/T$ ✓ $= -\dfrac{(+403\,000)\,J\,mol^{-1}}{773\,K} = -521\,J\,K^{-1}\,mol^{-1}$ ✓

🖉 You need to look up the values of the standard entropies of these three substances in the data booklet for (i) and the standard enthalpies of formation for (ii). Remember to use values for $SO_3(g)$ not $SO_3(s)$.

(iii) $\Delta S_{total} = \Delta S_{system} + \Delta S_{surr}$

$\qquad = +189 + (-521) = -332\,JK^{-1}\,mol^{-1}$

$\Delta S_{total} = RT \ln K$

$\ln K = \dfrac{\Delta S_{total}}{RT} = \dfrac{-332}{8.31 \times 773} - -0.0517$

$K = e^{-0.0517} = 0.950$ ✓

(iv) $\ln K = \dfrac{-332}{8.31 \times 1273} = -0.0314$

$K = 0.969$ ✓

(b)

```
                          ┌────────→ Ca²⁺(g)    +    SO₄²⁻(g)
                          │
         –lattice energy  │       ΔH_hydr|Ca²⁺    ΔH_hydr|SO₄²⁻
                          │              ↓               ↓
                          ↓
              CaSO₄(s) + aq  ──→  Ca²⁺(aq)  +    SO₄²⁻(aq)
```

Cycle ✓   Labels ✓

$\Delta H_{solution}$ = –lattice energy + sum of hydration energies of ions ✓

$\qquad = -(-2677) + (-1650) + (-1045) = -18\,kJ\,mol^{-1}$ ✓

(c) The value of $\Delta H_{solution}$ for barium sulfate is more endothermic by 28 kJ mol^{-1} ✓. This means that ΔS_{surr} becomes 28 000/298 = 94 J K^{-1} mol^{-1} more negative ✓. The value of ΔS_{system} of barium sulfate is 43 J K^{-1} mol^{-1} more positive ✓. As $\Delta S_{total} = \Delta S_{system} + \Delta S_{surr}$, the value of ΔS_{total} for barium sulfate is 94 – 43 = 51 J K^{-1} mol^{-1} less than that of strontium sulfate ✓. As ΔS_{total} is 51 J K^{-1} mol^{-1} less positive, barium sulfate will be less soluble ✓ than strontium sulfate.

The periodic table

Group

Key:

Relative atomic mass
Atomic symbol
name
Atomic (proton) number

Period	1	2												3	4	5	6	7	0
1	1.0 **H** hydrogen 1																		4.0 **He** helium 2
2	6.9 **Li** lithium 3	9.0 **Be** beryllium 4												10.8 **B** boron 5	12.0 **C** carbon 6	14.0 **N** nitrogen 7	16.0 **O** oxygen 8	19.0 **F** fluorine 9	20.2 **Ne** neon 10
3	23.0 **Na** sodium 11	24.3 **Mg** magnesium 12												27.0 **Al** aluminium 13	28.1 **Si** silicon 14	31.0 **P** phosphorus 15	32.1 **S** sulfur 16	35.5 **Cl** chlorine 17	39.9 **Ar** argon 18
4	39.1 **K** potassium 19	40.1 **Ca** calcium 20	45.0 **Sc** scandium 21	47.9 **Ti** titanium 22	50.9 **V** vanadium 23	52.0 **Cr** chromium 24	54.9 **Mn** manganese 25	55.8 **Fe** iron 26	58.9 **Co** cobalt 27	58.7 **Ni** nickel 28	63.5 **Cu** copper 29	65.4 **Zn** zinc 30		69.7 **Ga** gallium 31	72.6 **Ge** germanium 32	74.9 **As** arsenic 33	79.0 **Se** selenium 34	79.9 **Br** bromine 35	83.8 **Kr** krypton 36
5	85.5 **Rb** rubidium 37	87.6 **Sr** strontium 38	88.9 **Y** yttrium 39	91.2 **Zr** zirconium 40	92.9 **Nb** niobium 41	95.9 **Mo** molybdenum 42	[98] **Tc** technetium 43	101.1 **Ru** ruthenium 44	102.9 **Rh** rhodium 45	106.4 **Pd** palladium 46	107.9 **Ag** silver 47	112.4 **Cd** cadmium 48		114.8 **In** indium 49	118.7 **Sn** tin 50	121.8 **Sb** antimony 51	127.6 **Te** tellurium 52	126.9 **I** iodine 53	131.3 **Xe** xenon 54
6	132.9 **Cs** caesium 55	137.3 **Ba** barium 56	138.9 **La** lanthanum 57	178.5 **Hf** hafnium 72	180.9 **Ta** tantalum 73	183.8 **W** tungsten 74	186.2 **Re** rhenium 75	190.2 **Os** osmium 76	192.2 **Ir** iridium 77	195.1 **Pt** platinum 78	197.0 **Au** gold 79	200.6 **Hg** mercury 80		204.4 **Tl** thallium 81	207.2 **Pb** lead 82	209.0 **Bi** bismuth 83	[209] **Po** polonium 84	[210] **At** astatine 85	[222] **Rn** radon 86
7	[223] **Fr** francium 87	[226] **Ra** radium 88	[227] **Ac** actinium 89	[261] **Rf** rutherfordium 104	[262] **Db** dubnium 105	[266] **Sg** seaborgium 106	[264] **Bh** bohrium 107	[277] **Hs** hassium 108	[268] **Mt** meitnerium 109	[271] **Ds** darmstadtium 110	[272] **Rg** roengenium 111								

Elements with atomic numbers 112–116 have been reported but not fully authenticated

140.1 **Ce** cerium 58	140.9 **Pr** praseodymium 59	144.2 **Nd** neodymium 60	144.9 **Pm** promethium 61	150.4 **Sm** samarium 62	152.0 **Eu** europium 63	157.2 **Gd** gadolinium 64	158.9 **Tb** terbium 65	162.5 **Dy** dysprosium 66	164.9 **Ho** holmium 67	167.3 **Er** erbium 68	168.9 **Tm** thulium 69	173.0 **Yb** ytterbium 70	175.0 **Lu** lutetium 71
232 **Th** thorium 90	[231] **Pa** protactinium 91	238.1 **U** uranium 92	[237] **Np** neptunium 93	[242] **Pu** plutonium 94	[243] **Am** americium 95	[247] **Cm** curium 96	[245] **Bk** berkelium 97	[251] **Cf** californium 98	[254] **Es** einsteinium 99	[253] **Fm** fermium 100	[256] **Md** mendelevium 101	[254] **No** nodelium 102	[257] **Lr** lawrencium 103

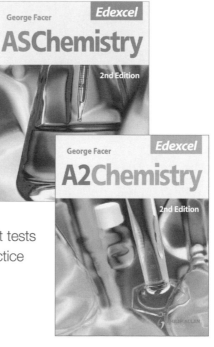